American Highway Roulette

Driving Risks and Consequences of Automobile Accidents

By
DENTON GAY

Copyright © 2010 Denton Gay
All Rights Reserved.

No part of this book can be reproduced, scanned, or distributed in any written or electronic form without permission.

All characters portrayed in this book are fictitious and any resemblance to real persons, living or deceased, is coincidental and unintentional.

ISBN: 1452876932
ISBN-13: 9781452876931
Library of Congress Control Number: 2010907197

Also by this author:
FATAL MISTAKES
TAILSPIN

This book is dedicated to my dear friend and first cousin, Dr. Richard Franklin "Dick" Smith, who died on March 9, 1980 as a result of injuries sustained in an automobile accident. I also dedicate this book to the millions of people whose lives have been tragically changed by this phenomenon that I refer to as *American Highway Roulette*.

PREFACE

The purpose of this book is to raise awareness of the dangerous risks and consequences of traffic accidents.

My own awareness began at an early age. Growing up near a two-lane highway in rural Arkansas, I lost several beloved animals to that narrow strip of asphalt.

In high school, the personalities of a couple of friends completely changed because of head injuries received in an accident. In another instance, a friend lost her life in a car wreck.

A decade later, tragedy struck again when a very close friend, who I thought was too smart and strong to die in an automobile accident, met his death from injuries he received in a vehicle rollover.

As an auto insurance claim representative and claim supervisor for several claims employees, I remained exposed to the experiences of those whose lives were changed by traffic accidents.

This history motivated me to write about ways to prevent traffic accidents and reduce the consequences that affect so many people. Many excellent articles have addressed individual aspects of automobile accidents. However, I am not aware of any treatment of "the big picture." Therefore, I feel my role is to present a broad perspective of this phenomenon.

My intent is to present the information in a way that is interesting and easy to read. However, some areas such as insurance and law contain terms and ideas that may be unfamiliar to some readers. Therefore, I included very simple reference sections entitled Common

Insurance Terms and Common Legal Terms for those who have not been exposed to those words.

Also included are chapters with descriptions of actual accidents. These are only examples and the personal information has been fictionalized. Because of the large numbers of accidents that occur in this country, some readers may think they recognize the identities of those characters. But almost every accident scenario has occurred many times over the past two decades. Only the basic facts in these examples are true, as reported in police reports and newspapers.

Ten percent of the profits from this book will be donated to charitable organizations promoting traffic safety.

Thank you for reading this book and a double thank you if you pass it on to others. I sincerely hope the information in it will help you and your families avoid the consequences of American Highway Roulette.

Sincerely,
Denton Gay, 2010

TABLE OF CONTENTS

Preface v

Chapter 1 - American Highway Roulette 1
    A) Auto accidents compared to roulette
        1. Statistics
        2. Analogy to war casualties
        3. Perception of accident risk
        4. Entities who profit from collisions
        5. Why consumers should facilitate positive change

Chapter 2 - Example 7
    A fictionalized version of an accident that actually occurred provides an example for use in understanding insurance and law.

Chapter 3 - The Insurance "Casino" 11
    A) A claim representative handles the claim from previous chapter
    B) Issues
        1. Claim values
        2. Employees of the Insurance "Casino"
        3. Cost of Admission
        4. Strengths of current system
        5. Weakness of current system
        6. What insurers can do to make the current system more effective and fair

Chapter 4 - Example 33
    A fictionalized accident illustrates problems for sensory-impaired drivers.

Chapter 5 - The Legal "Casino" 35
    A) Legal action needed for senior citizens and youthful drivers
    B) Accident victim navigates legal system
    C) Legal arenas
    D) Damage Issues

  E) Legal Issues
    1. Strengths of current system
    2. Weaknesses of current system
  F) How can the legal system be improved?

Chapter 6 – Example 53
  An accident reveals the myth of a "safe" car.

Chapter 7 - The Automobile Manufacturers "Casino" 57
  A) Automobile design changes
  B) Profits
  C) Motorcycles

Chapter 8 - Example 63
  A fictionalized version of an actual accident shows that immediate medical attention may not prevent death.

Chapter 9 - The Health Care "Casino" 65
  A) Typical experience with medical providers
  B) How can the health care system be improved?

Chapter 10 - Example 71
  A fictionalized accident involving teens illustrates need for action.

Chapter 11 - The Government "Casino" 73
  A) Reactive nature
  B) Government challenge

Chapter 12 - The Consumers "Casino" 79
  A) Reduction of accidents
  B) Minimizing consequences of accidents

Chapter 13 - Game Over 97
  A) Myths
  B) Truth

SUMMARY 101
APPENDIX - HELPFUL TOOLS 102

Safe Driving Tips 102
    - Ways to beat the odds

For Parents with Teens 107
    - Ideas for coping

Considerations in Buying Insurance 114
    - Price isn't everything

Common Insurance Terms 115
    - Words you should know

Common Legal Terms 117
    - Simple definitions

What to Do in the Event of an Accident 119
    - Avoid important mistakes

How to Resolve a Dispute 121
    - With or without legal representation

Buying a Car 124
    - Some safety considerations

Important Items To Keep In Your Car xxx
    - The life you save may be your own

Arkansas Law Regarding Wrongful Death Claims 129
    - An example for illustration

Potential Pitfalls In An Auto Insurance Policy Contract 133
    - Things to check to safeguard your protection

Sample Letter To Your Elected Representatives 143
    - An example you can alter

Web sites For Information 145
    - Good food for thought

CHAPTER 1

American Highway Roulette

Roulette provides recreation for gamblers who take chances with their money on the outcome of a spinning wheel in a casino. Would they play the game if their lives and well-being were on the line?

We engage in high-stakes risk daily on our highways. Spinning wheels transport us until our "number" comes up in the form of an accident.

In *American Highway Roulette* you expose yourself and others to a probability of death, mutilation, disfigurement, disability, and a multitude of other injuries each time you travel on a roadway. The only certainty in this "game" is financial because everyone pays.

Automobile accidents are the leading cause of death among people aged thirty-four and under. Yet, we drive unaware of the risks, much like the proverbial sheep that follow one after the other over a cliff. Most Americans accept the risk as a fact of life that remains outside their control. This inaccurate thinking serves to perpetuate the problem.

According to the United States Department of Transportation, National Highway Traffic Safety Administration (NHTSA):

> * **Over 6,000,000 accidents have occurred each year for the past sixteen years.**

* In 2007, 2,491,000 accidents involved injury.

* In 2007, 41,059 fatal injuries occurred.

In 2005, 43,510 human beings lost their lives as a result of *American Highway Roulette*. That averaged 119 fatalities each day or one every twelve minutes. In the United States, we've experienced more than forty thousand deaths from highway accidents each year during the prior sixteen years. The numbers have been fairly consistent from one year to the next.

Yet many government officials, business leaders, and others believe our highways are safe. They look at statistics per million miles traveled, and those numbers have declined over the years. They also compare our figures to third world countries.

Take a moment to contemplate the numbers given above. When you pause to consider how long we've been driving cars in large numbers, can you grasp the total number of people injured and killed in automobile accidents? In the twenty years preceding 2007, that translates to almost 40 million injuries and 800,000 deaths.

In 2008, we experienced a decrease in fatalities. The number listed by NHTSA is 37,261. This seems to be a trend because early estimates for 2009 are listed as 33,963 fatalities. More information on this trend is provided in chapter 12.

To put the deaths from automobile accidents in perspective, consider the casualties of war. How much time and attention do those fatalities receive in our media?

Probably not enough, but how much do you hear about auto accidents? Not much, by comparison. How much thought and conversation do we give present and past wars? Many books have been written, many movies have been made, and many hours of news media time have been devoted to cover wars. One example of the concern about war is the fact that hundreds of thousands of people demonstrated their opposition to the war in Iraq. Have you ever seen a demonstration for highway safety?

It seems we are blind to the tragedy that strikes us daily. Aren't the people dying on our highways just as important as those killed in war? If someone takes your life, does it really matter whether it was by an enemy's bomb or a stranger's car?

The number of Americans who have lost their lives in the Afghanistan and Iraq wars is fewer than half the number of human beings killed on the roadways of the United States each year.

If you die in battle, you may receive a medal of honor or be labeled a hero, and your family will be compensated. As this is being written, our legislators are considering increasing compensation for those killed or disabled while fighting in Iraq. Certainly, they deserve it.

As taxpayers, we all serve our country by working and supporting the actions of our government. Our tax dollars are used to pay our soldiers and the other costs of war. In a sense, couldn't the work we do be considered "in the line of duty?" For many of us, our duty is working to support ourselves and/or our families, employers, and government. While it is true that our level of service doesn't approach that of military personnel, we do contribute.

No honor comes from losing your life while playing *American Highway Roulette*. In fact, one moment of inattention could bring emotional, physical, or financial ruin to your family.

Think about it this way. If we were on a battlefield, would we place ourselves within a few feet of projectiles coming from the opposite direction? Of course not. But we routinely face two-thousand-pound masses of metal, traveling fifty feet per second, and carrying an explosive (gasoline). These deadly missiles pose danger every time we drive down a paved roadway, especially the two-lane variety.

Our thoughts are shaped by perspective. The families of those killed in the tragic events of 9/11 were offered generous amounts of money from a victims' fund. (According to a National Public Radio broadcast, at the time this was written, the average was $1.7 million.) Yet, if one dies in a car wreck, with few exceptions, only limited compensation is available from insurance or other private sources.

The 9/11 victims and their families received a great deal of attention from our media and politicians. In fact, a special commission was appointed to investigate and issue recommendations. A lot of money and time has been spent on this matter. While the attack was horrific and those affected deserve our sympathy, the point is, why aren't we having the same concern and dialogue about this greater menace? Highway fatalities, and their impact on families involved, are every bit as tragic. So, why don't we give this phenomenon more attention with the idea of making positive change?

Could it be that we are an arrogant people? Most of us think automobile accidents only happen to someone else. And if it does happen to you, the question that may form in your mind is, why me?

In terms of severe injury or death, some think, *I drive carefully, it won't happen to me.* Or, *I drive a safe car, so it won't happen to me.*" Maybe, *My children and I wear seat belts, so it won't happen to us.* The next myth is a favorite and epitomizes arrogance in our culture. *My reflexes are excellent, so I can drive fast without hurting anybody.*

These are just some of the falsehoods that people believe—until a tragic accident disrupts their lives, revealing the fallacy of their misunderstanding.

Another reason for coining the term *American Highway Roulette* has to do with the money involved. Just as casinos make a great deal of money as a result of their games, there are entities that profit greatly from automobile accidents. These businesses operate in several professions and industries. They all serve the needs of our society but could be improved for our protection and financial well-being.

One is the insurance "casino," where actuaries and underwriters calculate the odds. Agents and underwriters watch the door and determine both who can play and the cost of admission. Claim representatives serve as the cashiers.

The legal community is another beneficiary of our gambling vice. In their "casino," they make the rules; redistribute wealth, and right wrongs—at substantial profit.

The automobile "casino" represents the automobile industry, which benefits a great deal from our gambling addiction, by furnishing the "gaming machines" and their replacement parts.

The medical "casino" refers to the medical community. This high-stakes casino provides health care at a high price. But you don't mind because where else might your life might be saved or your pain lessened?

All these enterprises provide needed services for those who play the game. So, why should we be concerned?

Because we need to make changes for our own well-being. We cannot depend on those who benefit from the status quo to initiate change to it. The impetus for change must come through consumers' heightened awareness of the problems.

While the statistics for death and injury due to automobile accidents mentioned earlier are reason for concern, they are simply numbers—impersonal and devoid of image. Do they give you any *visual* picture or *visceral* feeling about injuries such as severe burns and conditions like quadriplegia? Think about your worst nightmare in terms of an injury. It can and does happen on a daily basis on and around our highways.

A friend of mine worked as an emergency medical technician (EMT) but decided to quit in order to maintain his sanity. Seeing a young person who had been decapitated in a car wreck was the event that compelled him to change jobs. He told me that the average length of service for the job was about two years. An Internet search revealed articles suggesting four years as an average while others mentioned ten. It appears that the geographic area and the specific type of EMS situation may make a difference. No definitive statistics could be found.

In order to provide an interesting and realistic view of *American Highway Roulette*, these chapters are interspersed with fictionalized accounts of actual accidents. The basic facts came from newspaper accounts. The individuals and relationships portrayed are imagined. Some may think they recognize those involved, but that is only because there are many similar accidents. The names and the factual information have been changed to protect the identities of those involved.

This is a tribute to those injured or killed simply because their number came up on the wheel of *American Highway Roulette*.

These victims aren't the only ones who lose in this version of roulette. Indirectly, those who pay insurance premiums and taxes bear the burden of financing these losses. That includes almost everyone.

Hopefully, the information and experience included here will motivate consumers to think about the problems described and decide what changes are in order. To understand why we should change the status quo, we'll need to look at the current system and explore the interrelated problems. Then, in later chapters, we'll look at the changes that could reduce the numbers and consequences of traffic accidents. Then you can ask the question, what can I do to make a difference?

Why should you care? Because nobody is immune from this terrible game of chance. All ages, races, sexes, religions, income levels, etc. are affected. *American Highway Roulette* is truly an equal opportunity event.

American Highway Roulette

CHAPTER 2

The Cramers lived the great American dream, which included a nice home with a mortgage, two late-model vehicles with payments, and two incomes. Today, they began a much-anticipated spring break vacation with their eight-year-old son, Bradley.

Jon, Christine, and Bradley carried suitcases out the front door of their home and locked it. Their green Ford Explorer and gray Honda Accord were parked in the driveway.

"Which car should we take?" Jon asked.

"The Explorer—it has more room," Christine said.

"Okay, we'll splurge. What's a little gas, anyway?" His eyebrow arched.

Christine stood by the driver's door with her hand outstretched.

"You don't trust my driving, do you?" Jon asked.

"You drive too fast." Her tone was matter-of-fact. "I'm not going to be a nervous wreck after a couple of hours on the road. Do you mind?"

"Heck no, Honey. I'll just relax and enjoy the scenery."

They climbed inside the Explorer. Bradley buckled himself in the back seat. Christine turned the ignition key, and the engine started. She glanced toward the back seat, checking Bradley's seat belt. "McDonald's, here we come."

Josh played with two plastic Power Ranger figures. "I'm hungry, Mommy. Dad said I could have a cim-ma-mon roll."

"Why sure, Hon. You can have anything you want. But there's no such thing as a cim-ma-mon roll. It's cinnamon roll."

She backed out the driveway and onto the street. After a quick stop at McDonald's, they began their trip to the amusement park, located just forty-five miles away.

Jon dug through a container full of CDs. "Hey, here's *The Best of Ray Stevens*. Bradley likes that." He inserted the disc into the stereo.

Christine followed the highway east through the northern foothills of the Boston Mountains listening to "Guitarzan." A long left curve loomed ahead.

A large Chevrolet pickup truck traveling west veered into their lane. Christine stomped on the brakes and swerved close to the right shoulder, causing the tires to make a short shriek.

In a split second all movement slowed, as if in the pulsating rhythm of a strobe light. The truck struck the Explorer head on in a resounding crash. The front end of the Explorer buckled inward as the airbags popped into view, partially obscuring the rising hood and image of the truck through the shattering windshield. Simultaneously, the deafening noise reverberated inside the passenger compartment. The air instantly filled with smoke-like powder from the airbags.

Christine's head turned instinctively away from the horror upon them and toward her husband in time to see his head moving forward. The plastic-covered metal on the right side of the windshield thrust toward him as his body moved forward. It struck his head, and he went limp in his seat belt and shoulder harness.

Bradley had been thrown forward and the back of the passenger seat pushed into his face. The vehicle rotated. For a moment it felt as if it might roll over on its right side.

"Oh my God!" she yelled.

The SUV came to rest facing oncoming traffic in the wrong lane, after being pushed almost 180 degrees. The sudden silence was interrupted by the screech of another set of tires as a third car tried to avoid striking their crippled vehicle.

Christine closed her eyes in prayer. A small black sports car swept past and into the ditch. She braced for impact. Moments went by before her shoulders relaxed. There would be no second collision.

Bradley cried from the back seat. "Ahhoow, ahhhhh, ahhhhhhh. My leg. It hurts, Mommy."

She looked back at Jon. Blood trickled from his nose, ears, and forehead. She burst into tears. "No, please God, no!"

A man rapped his knuckles on the glass of her door. "Are you okay ma'am?"

"No, we're not," she gasped and squinted to get a better look at him. "It's my husband. Please help him."

The man tried to open the door, but it was stuck. "Ma'am, can you check to be sure this door ain't locked? I can't get it open."

She pressed the unlock button. Nothing.

The man tried the back door. It opened. He stuck his head inside. "It's going to be okay," he assured Bradley and Christine. "I've already called the ambulance. They'll be here in a jiffy."

She turned back toward Jon.

He was dead.

CHAPTER 3

The Insurance Casino

A brown-haired, thirty-something male, dressed in tan Dockers and blue shirt with button-down collar, unlocked the glass door of the Good Heart Claims Office, one of several offices in a long, low-slung, strip center building. The dim light, empty parking lot, and silence testified to the early morning hour.

Inside, fluorescent lights revealed a typical business office—reception area, conference room, six desks surrounded by cubicles, filing cabinets, copiers, and a fax machine.

An older fellow looked up from a computer monitor on one of the desks as the man trudged past. "Morning, Bob."

"Mornin'," Bob mumbled and nodded. Somewhat pudgy, Bob positioned his bottom, widened by several years behind a desk, into its usual position. He logged onto the computer. "Geez. Six new claims."

"Seven, here," the older man replied.

The flat clack of Bob's computer keyboard signaled the beginning of another work day.

"You'd think we'd come in here just one day when nobody had a wreck."

Bob studied his computer screen. "Another freakin' fatality. Do you think that he would've left his house if he'd known his number was up?"

...

Bob, a fictitious insurance claims representative, will provide a glimpse into the world of automobile accident claims. As a person who financed his college education by working odd jobs, Bob knows the value of a dollar. Married and with a mortgage, he struggles to find financial success in the business world—like most young professionals in their thirties.

...

In the days leading up to Jon's funeral, Christine vacillated between raw emotion and total numbness. Yet, she had to help Bradley, deal with family, and carry out the dreaded chore of making final arrangements.

After several days, financial pressures demanded her attention. She reported the accident claim to her insurance company. A day later, the phone rang.

"Hello, Mrs. Cramer?"

"Yes?"

"This is Bob, with the Good Heart Insurance Company Claims Office. I'm very sorry to hear about your loss."

"Thank you." He seemed nice enough. Taking a deep breath, she felt ready to face whatever was to come.

"I'll be handling your insurance claim. If this is a bad time, we can talk later."

"No, I'll have to face this sometime, and it might as well be now." She summoned her courage and tried to focus on the conversation.

"Well, first thing we should do is talk about your insurance coverage. Our records indicate you have liability coverage with a $25,000 per person limit with a maximum of $50,000 for all liability claims."

His voice sounded matter-of-fact, but she wasn't sure what he meant.

After a moment of silence, Bob continued, "Liability coverage provides for any damages that you might be legally liable for to the other party as a result of the accident."

"Oh, but it was the other driver's fault."

"You also have $5,000 per person limit under the medical payments coverage. That pays medical and funeral expenses for anybody in your car." The words, spoken in a monotone, sounded flat and devoid of any feeling.

"I don't have any bills yet, but the funeral bill alone was over $6,000." Out of the corner of her eye, she noticed Bradley sitting in the living room, doing homework with the burdensome cast on his leg.

"Well, we'll wait until you get the bills, and then you can tell me how you want those benefits paid."

With a sigh, she wondered about paying the balance, the deductibles, copayments—and God only knew what else.

"Under that coverage, we do have the right to recover any amounts paid from the insurance company of the at-fault driver." Then he added, "That is, if his limits are high enough to cover your other claims."

This didn't make sense to her, but she wasn't sure how to phrase the question forming in her mind. The looming fear of money problems crept into her consciousness. Maybe the other driver's insurance would pay. "Will you be contacting the other insurance company?"

"Yes. Haven't they contacted you?"

"No. The police officer said the information would be on the police report, but that wouldn't be ready for seven to ten days." Her mind raced, thinking about the bills. She hadn't been back to work yet. Would the other insurance company pay for everything? When?

"You also have collision coverage, which pays for repair or the actual cash value if your car is a total loss."

"Oh, it's definitely ruined. The pickup truck hit us very hard." Images of the Explorer, on the day they had bought it, flashed into her head. Would they pay it off?

"Well, I'll take a look at it, and we'll go from there. Do you know where it's located?"

"No, somebody towed it away. I rode to the hospital in the ambulance with my son." The rising emotion from her chest formed a knot at the base of her throat. She glanced back at Bradley. He mustn't know about any money problems. There was already too much for him to deal with.

"It'll be on the police report. I'll pick up a copy in a couple of days." He paused, papers rustling in the background. "You also have a death benefit with a $5,000 limit."

Relieved at the unexpected news, she relaxed a bit. She hadn't known there was a life insurance provision on the car policy.

"You also have underinsured motorist coverage with a $25,000 limit per person and a $50,000 limit per accident. We won't know whether that will come into play until we determine the other carrier's limits."

Staring blankly out the window, she wondered what he meant. "How does that work?"

"Well, if the at-fault driver doesn't have enough liability insurance to cover your family's injury claims, then you could be entitled to benefits under this coverage you have with us. We'll just have to wait and see."

Silence followed as she tried to decipher and digest this information. "So, what's next?"

"I'll have to investigate the accident."

"Investigate?"

"Yes, I'll need to get your statement, the statements of anybody else involved, photograph the scene and the cars, and get the police report."

"Oh."

"In fact, would you be up to giving me a recorded interview about the accident today?"

Her internal alarm warned against it. She felt dizzy for a moment. "No. I'm thinking I should talk to my lawyer first."

"You have an attorney?" Without waiting for an answer, Bob said, "If you do, I'll need to work through him or her. Otherwise, it would be a violation of your attorney-client privilege."

"Not yet. But, I think I need advice." She paused again, "I've never been through anything like this before."

"Well, just be aware that an attorney will take one third of whatever you get. I'd like the opportunity to try to work with you." Bob concluded the conversation by giving her his number and asking her to call once she had made a decision about legal representation.

Her mind a whirlwind of thoughts and fears, Christine made a cup of coffee and tried to sort her problems out. She sat at the kitchen table with a pen and some paper.

Later she discussed the situation with her family and decided to handle the matter herself. After all, if things didn't go well, she could always hire an attorney later.

Meanwhile, Bob gathered information, inspected the car, requested a police report, and contacted other parties who witnessed the accident. He called back a couple days later and made an offer on the Explorer.

Christine chose to wait on the other company, so that no deductible would be charged. She also thought they should pay since it was their driver's fault.

A few days later the police report was available. Another few days passed before the insurance claim representative from the other company called. A week later that representative paid for the car and began working with Christine to get the medical bills. Several months would pass before all claims could be concluded. In some cases, it can be years before all claims are paid.

This is a typical sequence of events. Can a person traumatized by such an accident fully grasp all that is going on? What would you do in Christine's situation? To place this into perspective, you must look at the issues.

Obviously, this accident was tragic. A man's life ended prematurely. A husband is gone, and a father is dead. Imagine how this event changed the lives of the survivors.

Claim Values.

Our society uses money as a measure of value. Insurance is the current means with which we protect ourselves from financial loss. The victims in this accident have two basic places to get money: their own insurance company and the insurance company that provides liability protection to the at-fault party. Either way, the amount paid to Christine will be decided initially by an insurance claim representative.

This places the claim representative in the position of fairly compensating the accident victims while working for a business enterprise. In the wrong atmosphere, profit motives can influence insurance company management and employees. On the flip side, some can be prone to be overly empathetic or lack the assertiveness necessary to prevent overpayment of claims.

Please note that two claim representatives from two different insurance companies worked the claims arising from one accident. In regard to Christine, the adjuster for the at-fault party worked the third-party claims against his company's insured. (The liability claims are the amount the other party legally owes the Cramers. Christine becomes a third party to that insurer's agreement with its insured because its insured is legally liable for her damages.)

Bob, Christine's insurance company representative, handled her first-party claims. (In this example, the Cramers' first-party claims involve the amount their insurance company owes them on the basis of their insurance contract.)

If both parties are insured with the same company, one representative should be assigned to handle liability claims and a second to handle the contractual claims to prevent any conflict of interest. The reason has to do with the adversarial nature of liability claims, especially if liability is shared or disputed.

Both claim representatives must gather information and determine the amount to be paid. Usually, first-party claims will be paid as soon as they can be documented. The exceptions are uninsured and underinsured claims because they are handled in the same manner as liability claims. Liability claims are not paid until such time as all wounds have healed and a release can be obtained on behalf of the at-fault driver. If the injury claims are not substantial, and the liability limits are high enough to provide enough protection for the insured, the claim representative may elect not to take a release.

If a person has no insurance and a severe injury, the delay in handling any liability claims can create a hardship for that individual. For example, if the Cramers had no insurance coverage, they would have received no compensation until their liability claims could be settled—probably after Christine and Bradley's injuries had healed. Sometimes, that can take a year or more. Obviously, that would create a financial hardship for her.

If your household income was reduced by 60 percent and your expenses increased because of medical bills, how long would you want to wait? Some have suggested that our current system coerces the uninsured poor to settle claims prematurely.

In this example, we assumed the Cramers had insurance because most states now require all drivers to carry insurance. An insurance carrier would look at several issues.

How much are the Cramers' claims worth? What do you think a life is worth? What value for a good husband? A good father? A not-so-good husband or father? Immediately, you see difficulties emerge in trying to place values on human life.

Each state has legal guidelines to be used in evaluating death claims. Generally, you can claim monetary damages for sorrow, mental anguish, loss of companionship, loss of expected income, loss of services, and expenses incurred as a result of the accident. Most state law also specifies compensation for any conscious pain and suffering sustained by the deceased before death. The Arkansas provisions are listed in the appendix, along with the Arkansas Model Jury Instruction for a wrongful death claim.

How can we measure sorrow, mental anguish, and loss of companionship? We have no simple way to quantify these experiences. In order to be fair, most would say that the nature of the relationship between the affected family members is one aspect that should be considered. Another would be the age of the persons involved in comparison to their life expectancy. Some might argue that a fixed, flat value would be equitable. However, have you heard any intelligent discussion about these issues? Probably not—unless you sat on the jury of a civil trial.

What is fair value for conscious pain and suffering? This is another difficult question that has no correct answer. We have no way to measure the severity of pain.

Of course, one's life is the most valuable thing imaginable. We agree the survivors should be compensated, and it is easy to see that the problem lies in determining the dollar amount. Certainly, no sum of money can bring a loved one back. Still, as a civilized society, we should engage these matters with intelligent thought and discussion.

In the Cramer case, let's pretend that the magic number for the sorrow, anguish, and loss of companionship for their husband and father is $50,000.

The insurance representatives would evaluate the claim like this. Jon Cramer would be expected to live another forty years, and his net financial contribution to the family, after all taxes and living expenses, would have been $10,000 per year, given his current rate of pay. The recipients would be his wife and child, whose claims would be referred to as *derivative* because these claims derived from his injury.

His medical expenses were $2,047 and funeral expense was $6,300. Based on the above assumptions, the value of his claim is:

Mental anguish, etc.	$50,000
Lost income	$400,000 ($10,000 per year times forty)
Medical expense	$2,047
Funeral expense	$6,300
Total claim value	$458,347

How does that compare with your opinion of value?

Should we assume that Christine will remarry and reduce the amount allowed for income loss? One never knows. Had Jon lived and never been involved in an accident, he might have developed a debilitating disease—thus becoming a burden. This possibility is mentioned for the purpose of stimulating thought, not as a convincing argument.

Bradley had bruises, a broken leg, and minor cuts. His medical expenses were $7,500, and it took him six months to recover. He may have problems later in life because the bone in his leg had to be surgically bound together by mechanical devices. A report from his doctor, outlining the probability and cost of future surgery, would be important to any evaluation. The insurance claim representatives would likely evaluate it something like this:

Pain and suffering	$15,000
Medical expenses	$7,500
Future surgery	$6,500
Total claim value	$29,000

The amount for pain, suffering, and cost of future surgery could vary greatly, depending on circumstances, the doctors, and the claim representative. How much should Bradley's claim be worth, in your opinion?

Christine sustained bruises, two fractured ribs, and minor lacerations. She incurred roughly $1,500 for initial medical treatment and another $1,100 for psychological counseling. The insurance representatives might evaluate her claim like this:

Medical expense	$2,600
Pain and suffering	$5,500
Total claim value	$8,100

How much money should she receive for her injuries?

Opinions about the monetary value of a claim vary drastically. Try creating an accident scenario, and then get an opinion from your friends. The total dollar amount due the surviving Cramer family could range from one hundred thousand dollars to millions, depending on who does the evaluation.

As a former insurance company claim representative and later as a supervisor, I sometimes held meetings with claim representatives to see if the group could reach consensus on individual claim values. At other times, I participated in management group meetings where specific evaluations were discussed with the idea of establishing a value range for severe claims. The difference in opinions constantly surprised me.

The Cramers initially turned to the at-fault driver's insurance carrier for compensation. That driver carried the mandatory policy required by law. In Arkansas, most people buy the minimum required amount of liability coverage, which amounts to $25,000 per person with a total of $50,000 per accident. Is that enough to compensate a family in the Cramers' situation?

Worse yet, some states have even lower minimum requirements. See the comparison at the Insurance Information Institute Web site (www.iii.org). Click on the *Facts and Statistics* link, then on the *Issues Updates* link. Look for the *Compulsory Automobile/Uninsured Motorist* subject. Scroll down to the table that shows individual state requirements. Only four states have higher than the 25/50/25 minimum mentioned above. This Web site contains a wealth of information on insurance and highway safety topics.

Since state law determines insurance requirements, you can also look for information about statutes in your public library or on the Internet.

It would be enlightening to look at the wrongful death statutes in your state to discover the value of your life in the event you are an unlucky player in *American Highway Roulette*. If nothing else, it will help you determine how much additional insurance you need in order to take care of the financial needs of any surviving family, partners, or friends.

Consider the scenario of severe injury. What if one required total nursing care? At this time, the cheaper nursing homes cost $36,000 per year. Could your family support that? For how long? You think you'll "sue the bastard" who caused your injury? Better hope the person at fault has a lot of liability coverage; otherwise you are likely on your own because the sad fact is that most people have little in the way of financial resources.

In the Cramer case, the insurance claim representative for the at-fault driver would offer the maximum amount of his insurance policy ($50,000) for the settlement of the Cramers' liability claims. If the person at fault had personal property or money, perhaps the Cramers could sue them. But the truth is that most people live paycheck to paycheck, and assets are limited to equity in a home and maybe some money in a college or retirement account.

The insurance claim representative for the Cramers' insurer would pay the first-party claims applicable under their policy. The underinsured coverage would apply because the amount available under the liability coverage of the at-fault driver was inadequate. For the wrongful death claim of Jon Cramer, the insurance claim representative would offer the $25,000 per person limit.

For Bradley's claim, he would subtract the $25,000 paid by the other carrier and pay $4,000 difference, assuming he agreed with the $29,000 claim value.

On Christine's claim, he would pay the full $8,100 because the other carrier's limits had been exhausted. So, a good guess is that the surviving Cramers would be paid a total of $87,100 had they conducted a reasonable discussion with the insurance companies involved. The net amount would depend on the medical coverage available to pay medical expenses. Do you believe that to be adequate or fair compensation?

What if the medical expense had been greater? Would the claim be worth more? In most severe accidents, the medical bills would be many times the amount sustained in this example.

You can prevent financial catastrophe on a personal level by purchasing the maximum amount of liability, uninsured motorist, and underinsured motorist coverage, along with disability, health, life, and long-term care insurance. With such protection, you might be able to afford a catastrophic injury without ruining your family financially. But, you'll need a pretty good income to afford those insurance premiums.

Money cannot replace you, but it can help—especially if you are the breadwinner for your family.

Employees of the Insurance Casino.

It is easy to see how the insurance industry benefits directly from *American Highway Roulette*. Agents have the job of marketing, which in the casino analogy, means their job is to pack the house and collect the money.

They work with underwriters to determine who can enter the insurance casino and what it will cost to play. The underwriters do not want everyone to play in their casino. They want the people who are least likely to be involved in an accident.

Some players are considered uninsurable risks. They are such irresponsible, unlucky, and/or unskilled drivers that an insurance company cannot charge enough in premium to pay the claims generated. These people usually end up in an assigned risk pool where the state determines which company must insure them. They pay high premiums and almost always have the minimum amount of insurance coverage required by law.

The claim-handling employees function as payout cashiers in the insurance casino. They are usually salaried employees, although some companies outsource this function. They determine who should be paid and in what amounts. This involves deciding questions of coverage, investigating accidents, making liability assessments, determining loss values, and then negotiating and settling claims. Questions of coverage usually fall into two categories: whether insurance may have lapsed or whether either the car or driver was covered under the policy terms.

Cost of Admission.

According to the Insurance Information Institute, the average annual cost for auto insurance in 2004 was $838. In 2004, this varied by state from the lowest (Iowa) at $579 to the highest (New Jersey) at $1,221. Updated information is available at their Web site (www.iii.org). The actual cost also depends on the amount of insurance you purchase. The figures above are average, and those who carry comprehensive and collision coverage with low deductibles and large amounts of liability coverage pay more.

Agents earn somewhere between 8 to 15 percent of the premium, depending on the company and the duties of the agent.

A great deal of money flows into the insurance casino for redistribution.

Strengths of the present system.

Most insurers do a pretty good job. As business enterprises, they are very conscious of the costs involved and work to minimize them. Driven by competition from centralized operations with direct sales, several companies have reduced salary increases for employees and commissions for agents in order to keep premiums reasonable.

The industry invests premium dollars and uses the investment income to help maintain low rates. With investment instruments performing so poorly in recent years, this practice has not been as helpful as in the past.

The Insurance Institute for Highway Safety, an important organization in which most insurers participate, actively works to improve the safety of our cars and highways. Their Web site (www.iihs.org) provides a great deal of information. You can check the crash rating for many vehicles there. Of course, there are incentives for insurers to make cars and highways safer—it affects their profits.

If the insurance industry could simply pay claims per policy provisions and in accordance with a conservative view of applicable law, they could provide a good value for our insurance premium dollars—provided that no one disputed their decisions. Unfortunately, that is not the case.

Weaknesses of the present system.

Many problems cannot be controlled by the insurance industry. Why should you care whether the industry can control "their" costs? Because those costs are passed on to us, the premium-paying consumers.

In addition, some insurers make decisions based solely on money when a better decision might be based on what is right versus what is wrong or simply what makes sense.

Insurers spend a lot of money on advertising, promotion, and incentives such as conventions in exotic places. They also sponsor large sporting events. Since auto insurance is mandatory, should consumers' premium dollars pay these expenses?

Insurers have a difficult time enforcing policy provisions and controlling unreasonable claims. This is due to our litigious environment and the unfavorable image of insurance companies in this country.

For example, some insurers have tried to limit medical payments coverage to "reasonable and necessary" medical expense. Resulting lawsuits and bad publicity made that a very unprofitable venture. While a few people and their attorneys made some money in an attempt to punish these insurers, you and I picked up the tab as insurance customers.

Another example occurred when some companies tried to hold down repair costs by specifying aftermarket (generic) parts instead of original equipment (OEM) parts on car repair estimates. They were sued and several large verdicts were assessed against those insurers. If these large awards stand, they will be passed on to the public through insurance premiums.

In a similar vein, some companies list salvage parts on their repair estimates. For example, a fender from a salvage yard may be listed as an LKQ (like kind and quality) part. Usually, that will be an undamaged OEM part. If the paint is removed and the fender reprimed and painted to match, the repair should be acceptable. However, few, if any, will pay to do this. The result can be that the consumer will have a different color paint underneath, which can be very unsightly after a door ding or rock chip.

A third example of a problem in accident claims is diminished value. Most people believe that once an automobile is wrecked, its value has diminished, even if it is properly repaired. However, collision repair techniques enable competent body shops to restore a car to the point where the average person could not distinguish it from one that was never wrecked.

The theory behind diminished value is that one cannot sell the car for as much money if he or she discloses the vehicle has been wrecked. If we agree this is the case, and that one should be compensated, the question becomes how to establish such a value. Some have postulated that a percent of repair cost would be fair, while others believe it is the difference between wholesale and retail value. One argument for this position is that rebuilders (those who buy wrecked cars from salvage yards or auctions and repair them) usually sell cars at wholesale price. After considering the arguments for diminished value, is there a fair way to evaluate these claims?

Many hold the whole concept as wrong. Their argument is that, if the damaged parts are replaced with new parts and painted to match, what is the problem? Litigation regarding this issue gets different results in different states. Consequently, claims for diminished value will continue to drive up the cost of insurance coverage.

Another problem is insurance fraud. Fraud rings operate all over the United States. They stage accidents or use schemes to create accidents. Some insurers have inadvertently assisted these crooks by providing insurance bumper stickers for their customers. That tells the perpetrators that the car is insured. Some refer to these as *hit-me stickers*.

One tactic used is the "swoop-and-squat" accident. This is where a car cuts in front of another, and then the driver slams on the brakes. Sometimes, two or more cars are used to disguise the event. Often the damage is minor, but several people in the car will allege injuries.

This can result in having an accident on your record that may also cause your rates to increase. If your insurance company resists these claims, you will likely be inconvenienced with a lawsuit. The bottom line is that the rest of us will be paying the cost of the fraudulent accident.

According to the Insurance Research Council, fraud adds up to $6.3 billion of auto insurance premiums paid each year (www.ircweb.org). Insurers have made some progress in detecting and avoiding some of the consequences. But, all too often, insurers lack the necessary legal power to deal with these groups effectively.

Individual fraud cannot be calculated. After paying insurance premiums for years without presenting a claim, some people feel that when they do have an accident, it is their turn to "hit the jackpot." Many people feel that taking money from an insurance company is no worse than cheating on taxes owed the government.

During my experience as a claim representative and claim supervisor, many people exaggerated injury to obtain more money. This individual phenomenon of inflating claims is the huge iceberg beneath the visible tip identified as fraud.

According to information on the Insurance Research Council Web site, up to 30 percent of every claim dollar is lost to "soft" fraud. This is defined as small-time cheating by normally honest people. This individual greed can only be estimated, but is easily observed through personal experience.

Then, there is insurer fraud. This can occur at all levels within an insurance company. It can be a secretary who steals premiums from an agent, an agent who deposits premiums in his personal account, a claims adjuster who writes checks to accomplices, or an estimator who takes kickbacks from body shops or even acts of personal self interest by high-level executives. It is amazing, but not unusual because the people working in this industry are no different than those working in other industries. You can find interesting information about insurance fraud by visiting the Coalition Against Insurance Fraud Web site (www.insurancefraud.org).

In spite of the fact that most states require motorists to carry liability insurance, many people figure out ways to get around this. For example, they buy insurance to get car tags and then cancel. This irritated the taxpayers and premium payers of Louisiana, so they passed a "no pay-no play" law. If you don't have insurance, you cannot collect for damages, even if it isn't your fault. Their honesty about playing the insurance game, by including the word "play" in the title, is amusing. In the years since Louisiana enacted this law, several other states have enacted similar legislation.

Louisiana is a state where third-party claimants can sue the insurance company of an at-fault driver. A jury is more likely to award higher damages if they know an insurer will pay the judgment. In some jurisdictions, piracy has evolved into the legal arena, with judges and juries redistributing the wealth.

"Runaway" juries are a big problem in several areas of our country. It is impossible to get a fair and just result in these courts. The person bringing a lawsuit will win large awards almost every time. If you happen to cause an accident in one of these areas, you could most definitely be at risk of financial ruin—even if you have substantial liability limits. Certain counties in Mississippi, West Virginia, and Illinois are most notable examples of this phenomenon. Additional information about this subject can be accessed at Web sites listed in the appendix of this book. Insurers are unable to control this source of extreme costs.

Some state governments have enacted no-fault legislation to reduce these legally influenced insurance costs. No-fault systems are designed to provide compensatory benefits to consumers instead of trial attorneys. The pure no-fault concept removes the "blame" or fault component and only provides compensatory payments to consumers. However, by the time this type of statute gets through state legislatures, they are diluted to the point of being ineffectual. For example, according to an article on the insurance Web site mentioned earlier (www.ircweb.org), loopholes in New York's no-fault system cost their drivers an estimated $432,000,000 in 2003. Why do these loopholes exist in no-fault law? Because lawmakers, many of whom are attorneys, are influenced by members of the legal profession. Automobile accidents are a big source of income for many of their colleagues.

Insurers cannot control the legislative arena on other issues either. Several years ago, California lawmakers forced insurance companies to roll back insurance rates. That was an extreme example, but states pass new law each year that influences the amounts insurance companies can collect in premiums and the amounts they will pay out in claims.

Another wild card plagues the insurance casino. Their actuaries cannot predict astronomical jury verdicts that result from mistakes by employees involved in handling claims. One mistake in the "bad faith" arena can cost an insurer millions.

For example, in *State Farm* v. *Campbell*, the insurer was assessed 2.6 million dollars in compensatory damages and 145 million dollars in punitive damages. The Supreme Court ruled the punitive damages as excessive and sent the case back to the lower court with guidelines on keeping the punitive multiplier to a single digit number. In this case, the insurer was fortunate to get such a huge reduction. But, over a million dollars in excess of the policy limit was paid before the punitive damage award. And, the legal cost to defend the insurer must have been extremely high.

Although it appears that these types of awards are paid by the insurer named in the lawsuit, the reality is that they indirectly affect the insurance-premium-paying public. Most of these type cases are settled out of court. Those settlements go unreported by the media and unmentioned by insurance companies.

Another problem seen in some automobile liability insurance policies is that they pay punitive damages assessed against a policyholder by a third party. Usually, if excessive drinking of alcohol or other grossly negligent behavior causes an accident, the driver may be liable for these type damages. In our example given earlier, if the at-fault driver had been drunk, then the Cramers might have been able to collect punitive damages. However, if the at-fault driver's insurance company pays those damages under his liability coverage, the purpose is negated. How can the grossly negligent driver be punished when his insurer foots the bill? In the end, consumers bear the brunt of the "punishment."

We've seen how a big problem for those unlucky players of *American Highway Roulette* is that the average auto insurance policy does not offer enough protection. The required minimum amount of liability insurance is woefully inadequate to protect others for damages one might cause. How many cars cost more than twenty-five thousand dollars? What if you caused a two or three-car collision? One can easily be exposed to financial liability if the minimum limit has been selected.

Similarly, the first-party or contractual coverage you purchase to protect yourself can be inadequate. For example, in Arkansas, the standard auto disability coverage pays up to $140 per week. Would that pay anyone's bills if they could not work? For most people, it would hardly pay rent.

The standard death coverage pays $5,000, which doesn't last long in today's economy. An average funeral costs more than this coverage provides. One source advises the average funeral cost in the United States is eight thousand dollars, and another estimated total cost closer to ten thousand. This estimate included things such as burial plots, monuments, and other items not typically included in the price of a funeral service.

Standard medical payments coverage pays up to $5,000 for "reasonable and necessary" medical care provided to the insured as a result of an automobile accident. Some also provide that a portion can be applied toward funeral expenses. For any extended hospital stay this protection is the proverbial "drop in the bucket."

Therefore, automobile accidents cause the cost of health coverage to increase because most of the severe injury claims must be handled through health coverage insurers. If measures can be taken to reduce the injuries on our highways, it is probable that we could see a reduction in our health insurance premiums.

These figures are standard for an Arkansas auto insurance policy. In some states, the required minimum policy offers even less.

The attitude of most people is that each person should be responsible for his or her own well-being. If someone doesn't buy enough insurance and something bad happens, then it's his or her own "fault." But, how many people know the amounts their insurance policy will pay? Most think they have "full coverage" and that they are "covered" in the event of a loss. How much would your insurance policy pay?

If you can't answer that question, then have some empathy with the average person while giving serious thought to how much protection you really want for your family.

What about the family whose dominant member does not "believe" in insurance? Should the nondominant members be punished for their loved one's wrong thinking?

If one takes a broader view, don't we end up paying for those who aren't insured one way or another? The cost shows up directly in our uninsured and underinsured coverage. It also shows up in higher medical coverage costs, as medical providers have to write off a portion of debt owed by the uninsured. We also pay indirectly through our tax dollars when Medicare or Medicaid picks up the tab.

The insurance industry provides adequate protection for the average accident, but leaves the few severely injured at risk for loss. So, here's the problem. Most people involved in an accident will not be substantially affected but a few will find themselves in dire straits. Since only a few are affected in this way, no real attention is drawn to the problem. Most of those affected are unaware of what is going on with others, and simply accept their personal disaster and try to get on with their lives.

Perhaps the probability experts behind the insurance casino know the game, *American Highway Roulette*. If they covered all bets in our current system, the payouts might be so high that few could afford the premiums. Without changes in our current system, they may be right.

However, it is obvious that we are not willing to give up the game. We love our cars and the independence they provide. So, what are we to do? Will it simply be "tough luck" for those really unlucky *roulette* players?

Or, should we think through and discuss these issues with the goal of better use of our resources? We'll explore this in a later chapter after looking at the other interrelated problems. Each "casino" has a role in the overall problem and in the solution for minimization of the consequences resulting from our hazardous gambling addiction.

Statistics can be found at insurance company Web sites. Insurance is big business. If you are a member of the insurance community and wish to protect the status quo, I challenge you to create a plan to overcome the weaknesses of the current system as described above. If you are a concerned consumer, your challenge is to figure out what you can do to help them.

We do not know what the future holds. If something tragic were to happen to you or your family, the consequences may depend on what you do between now and then. If we can make positive changes, perhaps your life will be one of those saved through preventative measures.

If no changes are made, at least you now know that you should review your protection against a catastrophic injury and purchase the insurance you need to prevent financial ruin for your family.

What could insurers do to improve their effectiveness and fairness of their product?

They could focus on loss prevention. For example, they could offer law enforcement equipment such as video cameras or automated speed detection devices as tools for enforcement of law in regard to violations such as following too close and aggressive driving habits. A pilot program could be implemented to evaluate the effectiveness of this approach.

Consistency in claim handling could also be improved.

Insurers could also customize coverage. For example, I would not seek the services of a chiropractor for treatment of an accident-related injury. By charging me a premium for medical payments coverage that covers this type treatment, they collect premium for a service I would not use.

They could also offer alternative dispute resolution methods such as mediation, as a way to resolve disagreements. This could reduce the numbers of lawsuits and claims that currently require an attorney to resolve.

According to Congressman Morgan Carroll, the insurance industry spent $1,008,474,967 on lobbying from 1998-2007. His source is quoted as OpenSecrets.org.

The insurers could also cut costs in marketing, advertising, and administration. Or, they could remove those expenses in their accounting practices for mandatory products like auto insurance. Since we are required by law to purchase their product, do we owe these expenses?

Insurers could also become involved in education of their policyholders about the factors involved in the pricing of their product. They could support ideas and actions mentioned in chapter 12 to reduce the numbers and consequences of auto accidents.

From a consumer's point of view, the insurance industry faces the challenges of reducing accident frequency and severity in conjunction with providing an efficient and effective way to provide loss protection.

If a tragic accident strikes you or your family, would you be satisfied with the current compensation system? As a consumer, are you getting your money's worth? The costs of all the weaknesses listed in the current system are passed on to you, the premium-paying policyholder. The next chapters will show you where your money is going and how you can get greater value and protection.

American Highway Roulette

CHAPTER 4

Small, round, chrome hubcaps in the center of each wheel were the only shiny things about the old beige car sitting in the gravel driveway. A feeble, elderly man wearing a faded brown coat leaned against it.

"Getting old and being married to Edna Misner have one thing in common. They're a royal pain in the ass!" the spectacled man said, in a loud, hoarse whisper, as he opened the rear door to the Dodge Dart. His sparse gray hair shook while he struggled to place a smudged walker into the back seat. "Damn this thing." He began to wiggle the contraption back and forth to get it inside the car.

A thin, wrinkled woman in the passenger seat spoke hoarsely from beneath her scarf, "Tell *me* about it. It's a lot bigger pain *being* Edna Misner." She paused then spoke up. "My back is killing me, Teddy. Would you hurry up?"

"I'm hurrying as fast as I can." The old man staggered as he stuffed the last leg of the walker through the door. "You're just going to have to be patient." He closed the car door. The hunched figure was unsteady, keeping his hand against the car while stepping forward. "Always fussing." He opened the driver's door and practically fell inside. "Well, I'm in, but I'll be damned if I know whether I can get back out."

"Huh?" She turned a withered face toward him. "Did you say something?"

He readjusted the wobbly glasses on his nose. "I'm glad we don't have to do this more than once a week." His voice rasped. "If we get in any worse shape, I don't know what we're going to do."

His hand trembled as he tried to fit the key into the ignition. "I don't know why they had to make these cars where you can't see the ignition switch. I liked 'em better when they were right in front of you." Purple veins stood out under the opaque skin of his hand as his finger tapped the dash.

"I'm cold. Turn on the heater." The delicate woman shivered slightly, even though she wore a heavy coat.

Several long seconds later, he thrust the key into the ignition switch and started the car.

"What time is your appointment with the doctor?"

"It's at ten forty-five, but if we get there earlier, maybe I can get in. I'm hurtin', bad."

Ted steered the car onto the highway and began the trip to town. Rain began to fall and the worn-out windshield wipers did a clickety-clack squeak against the windshield. He strained to see through the rain-streaked window. With both hands gripping the wheel, he drove slowly along the gray strip of pavement. Raindrops splattered against the car, creating a metallic sound inside.

His expression never changed as he steered their car across the faint, double-yellow lines in the middle of the road. The oncoming truck looked surreal, as if it had just magically appeared from the grayness of the downpour.

The sound of the crash was deafening. Both Ted and Edna slammed against the spider-webbed windshield, their heads leaving inverted bubbles there, after their lifeless bodies crumpled limply, like rag dolls, against the dashboard and back into the seats.

Their suffering ended suddenly, but for the occupants of the pickup truck, it was only the beginning.

CHAPTER 5

The Legal Casino

The situation in the last chapter raises the question as to whether there is some age when drivers should be retested or barred from driving. The demographics of our country indicate there will be a huge influx of these type drivers as the baby boomer generation ages.

One estimate indicates that by 2025 there will be thirty-three million drivers over seventy years of age. Restricting or testing older drivers meets resistance from senior citizens. Some are very good drivers at age seventy-five, while others have difficulty at sixty. Few senior citizens are willing to give up their right to drive because it restricts their freedom or renders them dependent. Nevertheless, we need to address this problem and create law that protects everyone.

The crux of this issue is whether a person has sensory impairment. As we age, changes in the body occur normally. These can result in hearing loss, decreased depth perception and other vision problems, slower reflexes, loss of brain function, inadequate range of motion, and other problems which directly affect driving skills.

According to a study by the AAA Foundation, drivers sixty-five or older are nearly twice as likely to die in a crash as drivers between fifty-five and sixty-four. Drivers over eighty-five were nearly four times as likely to die because of the increased fragility of older people. This study also concluded the higher probabilities were due to lapses in perception. Based on the fact that we'll soon have a large influx of such drivers, steps should be taken now to prevent these type injuries and deaths.

One Missouri couple took action after an incompetent elderly driver killed their son. For the full story, please visit their Web site, Concerned Americans for Responsible Driving, Inc. (www.drivingsafe.org).

After spending considerable time and money, they persuaded state legislators to pass a law whereby a family member, physician, or others could anonymously report an impaired driver believed to be at risk. That driver would be retested for safety purposes.

Compared to mandatory testing of all drivers, the benefits of this method are twofold. Not all senior drivers would have to be tested, and it allows familial relationships to remain intact that might otherwise be irreparably damaged.

One disadvantage is that some drivers may not be reported until it's too late. However, such a law might have prevented the accident described in the last chapter.

Similarly, the situation with young drivers needs serious thought, debate, and legislative action. It's no secret that young, inexperienced drivers have more accidents. That is why the insurance premium rates for them are so high.

According to the NHTSA Web site, the sixteen to twenty-four-year-old age group constitutes 24 percent of all traffic fatalities. The need for action in this arena is clear. Most of us have personal experience to draw on in thinking about this problem.

You can review recommendations and see what legislation various states propose, by visiting the Web sites listed in the appendix of this book. Most model legislation deals with graduated licensing provisions with elements as follows:

Stage 1: Learner's Permit
Minimum age requirement
Pass vision and "rules-of-the-road" knowledge tests
Demonstrate basic vehicle operation skills
Must be accompanied by a licensed adult at all times
Require specified number of hours of supervised practice

All occupants must wear safety belts at all times
No alcohol or drugs while driving
Differentiated permit
Cause no "at-fault" crashes and have no traffic offenses for six months

Stage 2: Intermediate License
Complete stage one
Meet minimum age requirement
Pass a driver's test
Complete advanced driver education training
All occupants wear safety belts
Nighttime driving restrictions
Passenger restrictions
No alcohol or drugs while driving
Driver improvement actions are initiated at a lower point level than for adult drivers
Intermediate license has markedly different appearance than regular license
No "at-fault" crashes or traffic violations for at least twelve months

Stage 3: Full License
Complete stage two
Minimum age requirement
No alcohol or drugs while driving

    Recently, there have been movements to raise the legal driving age to eighteen-years-old. To be sure, most teens will not want to see this type law enacted. Some teens are more mature than others but all need more extensive training and monitoring than they currently receive.
    Speaking of law, let's imagine that Christine Cramer from chapter two decided to seek legal help, as an example of how our current legal system works. After all, she was overwhelmed by the complexities of insurance.
    Browsing through the many legal advertisements in the telephone directory, she elected to trust her intuition and picked an honest-looking face. It belonged to an attorney named Preston.

His receptionist took her call, obtained a bit of information from her, and then set a date for the free consultation.

After several days of receiving letters from other attorneys soliciting her business, she met with Preston in his law firm's office. An elegant reception area led to a plush office with bookshelves full of leather-bound law books. An older version of the photograph in the telephone directory greeted her as she was escorted into the office.

"Hello. I'm Preston." A middle-aged man with tints of gray in otherwise brown hair stood from behind his desk and extended his hand.

She accepted his handshake. "I'm Christine Cramer."

"Please have a seat," he said, waving to a chair. "I'm sorry to learn about your recent loss."

"Thank you." She sat in front of him. "I need some advice. I've never been through anything like this."

"I understand." Smiling, he sat. "I'll be happy to help you in any way that I can."

"Well, I'm just not sure. The insurance man called me and started explaining things but..."

"Insurance can be very confusing. Have you signed any documents?" He toyed with a pen as he spoke.

"No. But the person I talked to wanted me to give him a recorded statement. Do I have to do that?"

"Absolutely not." He leaned forward. "We don't recommend it. I've seen insurance adjusters place words in a person's mouth, mislead them, and twist statements around to make them say the very opposite of what was intended."

"Well, that's my concern." She thought back to the conversation about policy limits, and then added, "And, I'm worried about money."

"Insurance companies are businesses," He leaned back in his chair. "And those adjusters are going to save as much money as they can for the company. That's what they get paid to do." Index finger pointed at her, he continued. "That's why you need a lawyer."

Although she nodded her head, her decision wasn't yet made.

He studied her reaction. "I can guarantee you that if you hire me to represent you, I'll get every dollar that's coming to you."

That idea appealed to her, but..."How much do you charge?"

"We don't charge you anything. We take our fee from the insurance company. If they don't pay what's coming to you, then we'll take them to court. If we lose, we don't charge you a thing."

"That sounds pretty good."

Preston picked up the phone and pressed an intercom button. "Please bring me our standard fee agreement."

A secretary brought the document into the office.

"We take a one-third fee out of the settlement. Now, if we take the case to court, we'll have to charge more for the extra work, so the fee could go up to one-half."

Raising her eyebrows, she wondered if he had actually said what she'd heard—half.

"But, we rarely have to go to court." Wrinkle lines formed across his forehead as he raised his eyebrows. "Believe me, the insurance companies know this firm. They don't want to face us in court. We settle 95 percent of all our cases."

She squirmed, trying to fully grasp the implications.

"And, you have the final say about whether to settle or try the case. So, if the amount isn't satisfactory, we'll take 'em to court."

The words reassured her. As long as she had control, what could go wrong? If she could just get enough to replace Jon's income, pay the bills, and maybe fund part of Bradley's education.

"You know, law has become very specialized. Some firms do tax law, some divorce law, but our firm specializes in auto accidents. We do this every day. I'd like to take your case."

Christine paused, trying to sort out her decision. What else could she do? "Okay… sure."

He smiled at her with confidence.

"Great." He pushed the paper toward her. "Take your time and read this agreement. It's our standard form. Just sign right there when you're ready."

She glanced over the agreement.

"Any questions?"

"No." She signed the document.

"My secretary will give you a form to fill out about the accident. Once we get that, we'll go to work for you."

Relieved, she rose, thanked him, and walked to the reception area.

Preston stuck his head out of his office. "We'll also need photos of you and your son."

"Okay." Christine smiled. After filling out the form, she set an appointment to come back for the photos. Then she left the law office, relieved that she wouldn't have to face the ordeal alone.

Once she arrived home, she called the insurance representative, Bob, and told him that she had decided to use an attorney to handle her claims. Bob advised that he would work directly with the attorney.

A few days later, the at-fault insurance company representative called. Christine felt grateful that she could tell that person that Preston, her attorney, would be handling the case.

Bob also contacted Preston and made an offer on the car. But, the other company had been in touch with Preston, and he decided to have them pay for the Cramers' car. That way, there would be no deductible.

The next conversation between Bob and Preston would go something like this.

"Preston, this is Bob with Good Heart Insurance Company."

"Yeah, Bob, how are you?"

"I'm calling about the Cramers' claims," Bob said.

"Did you get my limits demand?"

"Yeah, but a demand for policy limits seems premature. Do we have all the bills?"

Preston pulled a file from his desk and flipped it open. "I'm showing $8,347 in medical and funeral expenses for Jon, $2,600 in medical expense for Christine, and $7,500 in medical expense for Bradley. Is that what you have?"

"Yeah. But do you think that warrants limits?"

"Look, Bob, the other carrier has fifty thousand dollar liability limits. The Cramers have a death claim and over eighteen thousand in bills."

"Yeah, but you've got to look at the per person limits. Sure, Jon's death claim is worth the twenty-five thousand limit of the liability carrier and the twenty-five thousand limit of our underinsured coverage. But, are Christine and Bradley's claims worth the remaining twenty-five thousand of their liability coverage plus the remaining twenty-five thousand under our underinsured motorist coverage?"

"Geez, Bob. These people are losing big here. Jon's claim is worth a million dollars, but there's only fifty grand available. I think you can be a bit liberal on the wife and son's claim."

"Well, I call 'em like I see 'em. I'll sit down and evaluate them in the next day or so and get back with you."

That concluded their conversation.

Although they might argue about whether the mother and son's claims were worth the remaining limits, Preston would likely prevail. However, that begs the question of whether Christine would have been able to persuade Bob of that value, had she tried to handle the claims on her own.

At any rate, the last conversation that the attorney would have with Christine would go something like this.

"Here are the checks from both insurance companies. Please endorse them."

Christine looked at the checks. Each was for $50,000 and had the attorney's name on them. One even had her insurance carrier on it. "That doesn't seem like much for everything we've been through."

"I got the maximum amount available. It's too bad. The case was worth a lot more."

"Why is my insurance company's name on this check?"

"Well, they have a right to recover for the medical expenses they paid, but in this situation they elected to waive that right. Fact is, they really had to because you hadn't been adequately compensated." Preston handed her a check from his trust fund.

Christine studied it carefully. Sixty-six thousand dollars, a lot of money. She wondered how long it would last. Would she and Bradley be able to keep their house?

One also wonders whether the insurance company would have waived their right to recover the amount of their medical payments had Christine attempted to handle claims directly with them, without a lawyer.

Just like the Cramers, you cannot depend on obtaining the legal value of your claim because of the inherent problem discussed in chapter 3, which is money. The Cramer family income had been cut by 60 percent. Even before the accident, the dual-income family struggled financially.

Did you notice that Preston's one-third fee came to $33,333 (1/3 of $100,000) which is exactly half as much as Christine and Bradley received? Behold the beauty of math.

Welcome to the "legal casino." Law is written in conjunction with legislators and evolves within our legal system. Granted, a good number of legislators are not lawyers, but many are and those who are not depend on attorneys to properly write the law. Automobile accidents are a boon to the legal profession. In addition to the legislative aspect, there are two legal arenas applicable.

Criminal law enforces the statutes written to protect people while using the highway. This is where traffic citations or tickets come into play. If you speed, you are cited for breaking the law. A judge assesses a fine and you pay the court.

Some attorneys become judges while others become prosecutors or specialize in defending criminal cases. Issuing tickets, prosecuting tickets, and paying fines are not popular activities.

The second legal arena is civil law. This is where the legal casino keeps a lot of money. Most people are not aware because they are not directly affected. This is the component in our legal system that deals with damages as a result of another's negligence. Here, financial remedy is provided when one person wrongs another. In other words, this is the place where Preston, the plaintiff attorney, and his clients (accident victims) can get money for injuries or death.

Automobile insurance companies usually handle these legal liability claims. Let us look at an example of the typical automobile accident and how it works through our legal system.

Imagine that I am driving down the road, minding my own business, and stop for a line of traffic in front of me. You follow me in your car but become distracted and cannot stop before colliding with my vehicle. This is the most common type accident in the United States today.

A police officer investigates and gives you a ticket for following too closely. We exchange insurance information, and if the cars are drivable, we go about our business. You will likely not see me again.

You will go to traffic court or forfeit bond and pay the court a fee for your violation of the law.

There will be contact between your insurance company and me. If you are insured by a competent company, they will try to contact me and handle my claim. They will pay for repairs to my car (assuming the damage is economically repairable) and possibly pay for a rental car while my car is repaired. They will also evaluate and pay my injury claim.

To keep this example typical, we'll pretend I have the common whiplash injury. If your insurer and I agree on all these values, they will pay me, obtain a release on your behalf and the matter will be closed.

If there is a dispute, I will hire a plaintiff attorney to pursue my claim against you. If the dispute cannot be resolved, he or she will file a civil lawsuit against you, on my behalf.

The defense attorneys hired by your insurer would respond on your behalf and are generally paid by your insurer per the provisions of your liability insurance coverage. Defense attorneys charge one hundred dollars per hour and upward, depending on where you live.

The plaintiff attorney usually has a one-third fee if the case settles. The fee increases up to one-half if the case goes to court and must be tried to conclusion. There are some variations of this fee agreement within the legal community.

Let's look at what happens if your insurance company and I disagree on the value of my claim. What's a claim worth? It is a matter of opinion. Rarely do two people (independently) arrive at the same amount. Two insurance claim representatives (also referred to as adjusters) would likely place differing values on a claim. Similarly, two lawyers, two judges, and even two juries would not share exact views. But to give you an answer to the question of what a claim is worth, one plaintiff lawyer explained it this way. A claim is worth what you can persuade a claim representative it is worth, or, it is the value that a jury of twelve will give it in a court of law. That is the truth.

Damage issues.

Most property damages can be assessed and argued with concrete evidence. Most body shops are on very similar labor rates and parts are generally consistently priced. There can be arguments over the value of a car, but advertisements and sales records can be used to support a realistic conclusion.

Generally speaking, property-damage issues don't play a major role with the legal profession. There are exceptions and some big ones (generic parts versus OEM parts, diminished value, etc.) mentioned in an earlier chapter.

The injury claim is the most interesting part of liability claims. In our example, we'll assume that I live in a conservative area and that I am a reasonable person with this typical "whiplash" injury. I'll visit my family doctor, who will diagnose me with a cervical strain. He'll prescribe a muscle relaxer or anti-inflammatory medicine and possibly pain medication to help my discomfort. Depending on the pain and my threshold for tolerating it, there might be some physical therapy prescribed. With conservative treatment, my expenses may only be six or seven hundred dollars. It may take four to six months before I'm feeling back to normal. However, I may feel fine in two weeks. At any rate, we'll say it has been six months, and I have all my bills, which total seven hundred dollars. What is this claim worth?

An insurance company or attorney will look at it like this. One is entitled to reasonable medical expense, reasonable income loss, and compensation for pain and suffering. How does one attach a dollar amount to pain and suffering?

Somewhere along the way, some enterprising attorney or insurance adjuster decided that figure must be based on the amount of medical treatment. The logic was that more medical expense meant a more severe injury, and therefore greater pain and suffering. So, they placed an arbitrary multiplier on the medical expenses. Though some injuries, such as a broken bone, clearly nullify this method of evaluation, some attorneys and insurance representatives in this business agree that three times reasonable medical expense could be a fair way to determine this compensation. However, I've also heard some attorneys claim that five to ten times the medical expense determines what a case is worth. Some insurers and attorneys will reject the whole notion of these arbitrary multipliers.

For the sake of this example, let's assume that three times the amount of my medical expense is fair. Three times seven hundred gives a value of $2,100 for pain and suffering. Therefore, according to some experts, $2,800 is the value of the case. So, I make a demand (offer) to settle for that amount.

Your company argues that the total amount of the claim is $2,100 because they evaluate the three times medical as all-inclusive. They also point out that there wasn't a lot of damage to my car. They refuse to pay what I have asked and stand on their $2,100 offer.

So, I consult my attorney, who says he can get more for the case. Now, I have to make a decision. Do I hire the lawyer? If he wins, I may break even or maybe come out a few dollars ahead. If he can't get a verdict or settlement above the offer large enough to cover his fee, I'll lose. But I'm thinking that the offer is low and that I'd rather my money go my lawyer friend than to some heartless insurance company. I hire him. He contacts the insurer but cannot get an increase in the offer. So, he files a civil lawsuit.

You get a summons and complaint in the mail or perhaps you may be served in person by a sheriff or process server. You give it to your insurance company, and they hire a defense lawyer. The defense lawyer enters a plea and legal discovery begins.

After each side answers interrogatories, the attorneys take depositions and file motions, depending on the nature of the case. In my case, they'll obtain medical records, and then likely depose me and any doctor who has treated me. There will be negotiation at various stages of the case. If those negotiations fail, a year or so after the suit is filed, there will be a trial.

The outcome of the case will depend on the place where it is tried. Each state has different laws regarding automobile accidents. The attitudes and values of juries vary from place to place. One cannot anticipate the outcome without knowing something about the jury hearing the case. This can be very important to you because a liberal or antagonistic jury could award damages in excess of the coverage afforded by your insurance policy. In other words, a mistake on the highway by you or your children could cause you to lose money and/or property.

For the sake of discussion, let's say that I did better than average (in my conservative venue) for such a minor claim and received a verdict of thirty-five hundred dollars. Because the case went to trial, my attorney would receive 50 percent. That amounts to $1,750, leaving me with the same amount. Subtract my $700 medical expense for a $1,050 net outcome. I'd also have some court costs to pay. My attorney would end up with more money than me. In this example, I'd have been better off taking the offer from the insurance claim representative.

Now, let's look at the larger picture. Your insurance company would pay somewhere between five and seven thousand dollars for defense of the case, depending on what issues arose during discovery, plus the $3,500 judgment. The total amount spent on this conservative, common claim would be between eight and eleven thousand dollars—where the actual loss was less than one thousand. Can you see the impact of the legal casino? Most of the dollars go to the dealers.

Actually, relatively few cases go to trial. The typical claim is settled out of court. In the claim I am describing, a plaintiff attorney will usually write a few letters, make a couple of phone calls, and negotiate a settlement in the $3,500 to $7,500 range.

Why that range? Do the math. It becomes worth it to defend a case that involves more money. Why resist a case that costs more to defend than pay? It makes immediate economic sense.

However, some insurance companies have resisted paying these type claims. Because people are playing this injury claim game without being truly injured, some in the industry believe that it is better to incur legal costs than let this avarice go unchecked. This thinking is geared toward the long-term situation. So far, this strategy has not made an observable difference.

The average cost of an injury claim continues to rise. The most recent information available indicated an average cost of five thousand three hundred dollars. That does not include legal fees or any other expenses associated with the claim.

The example used for our accident was a conservative one. Many people would seek additional treatment and have three to five thousand dollars in medical expense. Then a settlement would be proportionally higher. The attorney would likely make three or four thousand dollars for a few hours work.

In one severe injury claim, an attorney can make a fortune. One-third of a couple million is a tidy sum of money. In a case where twenty-one children were killed in a school bus-beverage truck accident, the beverage truck insurer quickly agreed to pay $122 million to settle the claims. One expert estimated that the lawyers for the families received the equivalent of $25,000 an hour for their services.

While that would be a most extreme case, can you see why there is so much advertisement by plaintiff attorneys? Look in the yellow pages of your phone book or watch television. People involved in auto accidents routinely receive letters, videos, and phone calls. "Runners" for plaintiff lawyers obtain police reports on a daily basis for purposes of soliciting auto accident victims' business.

Insurance companies, doctors, and business leaders often blame our legal system for driving up costs. There are many frivolous lawsuits and there are many zero-dollar verdicts, especially in regard to automobile accidents.

Average jury verdicts vary, depending on who does the research, and they can appear very high or very low. A more accurate way to look at them is on a median basis, which tells you the amount of most verdicts. Most estimates for median verdicts resulting from automobile liability claims are around eighteen thousand dollars. This could be viewed as evidence that our legal system works well in most cases. Those do not typically get media attention. Only the ridiculous awards make the news. Still, when you consider that most judgments and defense costs are paid through some type insurance, guess who indirectly pays the bill? We do, whether through higher premiums or the cost of goods and services.

Consider the MADD (Mothers Against Drunk Driving) example. If you lost a child because of a drunk driver, you would be angry, too. MADD's efforts to fight drunk driving deserve commendation. You'll notice that some plaintiff lawyers see the opportunities there and have jumped on the bandwagon, hoping to snag a big case where punitive damages may apply.

In fairness, one should consider that attorneys, like other professionals and businesspersons, have overhead and other costs. Their fees are not pure profit. Also, please note that many attorneys would not use language included in the Christine Cramer example. That language was selected based on advertising by some of the more aggressive plaintiff attorneys.

Legal issues.

Consider the legislative area of our legal system. Currently, we have many attorneys and legislators making law in each state. Do we really need that many sets of rules? Couldn't we all live by one? If one federal group of statutes defined the law for all states it would simplify our system, making it more effective and efficient.

For example, in Arkansas, we have a *less than fifty* comparative negligence rule. That means one must be less than 50 percent at fault in order to collect damages from another driver. If liability is assessed to me at 30 percent, then I can collect 70 percent of my damages. In other states with pure comparative negligence statutes, I could collect something, even if I was 50 percent at fault. In Arkansas, accidents deemed fifty-fifty result in nobody getting paid. In my opinion, this has been used by some insurance companies to avoid paying questionable claims.

The theory of negligence underpins the civil justice system. Does it really serve us? For example, I've driven for many years without an accident. If I look away momentarily and rear-end a car in front of me, am I negligent? One moment of inattention out of countless hours behind the wheel would not seem to constitute negligence. People make mistakes and our current tort system does not serve to prevent accidents.

Other problems with our current system need scrutiny. Joint and several liability law can be a problem. As an example, pretend you were driving five miles an hour over the speed limit when a car pulls on to the road in front of you. There is no time or space to stop and a collision occurs. The driver and passenger of that car sustain severe injuries. Because there is no negligence on the passenger, he or she can sue the at-fault driver and you. If the passenger obtains a large judgment, you could be assessed with a big part of the damages, even though your negligence was far less than the other driver. This has happened.

Many of the people who make our law are attorneys. Most are fine, caring individuals. But don't expect our lawmakers to make changes that create a reduction of income for their brethren. Politics dictate their decisions. The point is that you will not see the legal profession, in general, take great measures to make the highway a safer place to drive. They will represent accident victims and attempt to get top dollar for those claims, but they don't want substantial change. Can you blame them?

According to information on the Institute for Legal Reform Web site (www.instituteforlegalreform.com), the tort system in America costs businesses and consumers $261 billion annually. Those figures include more than just auto accidents. The 2006 statistics translated to $880 per citizen.

Strengths of the current system.
In the majority of cases, the current system provides a fair opportunity for an individual, or group of individuals, to obtain justice. In most cases where a plaintiff has no legitimate claim, he or she loses. In most cases where a valid case is tried, the plaintiff is awarded appropriately.

Weaknesses of the current system.
The greatest criticism is that the legal process is too expensive. Relatively speaking, not enough money goes to compensate the victims.

It is also too complex because each state has different law spelling out the rules. A lot of time and money are spent in state legislatures when one national system would promote equity, fairness, and simplicity.

A third weakness is that results are too inconsistent. Is it fair for everyone to be compensated differently for the same type injury?

A fourth weakness is that it is too time consuming. The delay in compensation can create hardship for some people. Delay can be extensive, especially if a verdict is appealed to a higher court.

What could the current legal system do to make our system more efficient and effective?
Utilize technology to streamline the legal process and then pass the savings on to consumers. In years past, lawyers would have to physically travel to the courthouse to file legal documents. Often, preparation of the paperwork took a great deal of time. If both these time-consuming activities can be significantly reduced by modern technology then more dollars could be placed into the hands of accident victims.

Lawmakers could also work toward one set of statutes for our country. Simplification could save more time and money.

A jury selection system must be developed that eliminates bias and unfair jury verdicts. Perhaps jury pools could be composed of knowledgeable people, as opposed to random peers.

Efforts could be channeled toward getting more money to injured persons, instead of preserving the status quo. One way to do this would be to place a cap on contingency fees.

From the consumer point of view, we want an effective and an efficient system for fair and equitable dispute resolution. This is the challenge the legal profession must meet in order to maintain the status quo.

If you are a member of the legal community, we, the consumers, ask you to come up with ways to overcome the weaknesses in the system as described above. Many of the legal procedures used today have been around for decades with little or no change. Couldn't the system evolve for greater efficiency and effectiveness?

American Highway Roulette provides big business to the legal profession. Do you want to play in their casino? You and I, the consumers, pick up the tab. That cost can be reduced.

Alternatives presented in chapter 12 should be considered, discussed, and solutions implemented. First, let's take a look at some other, interrelated problems.

American Highway Roulette

CHAPTER 6

A gray-haired woman in her late fifties made coffee in a dimly lit kitchen while listening to a morning news show on a small television. A velour robe covered her plump body. She poured the water into the coffeemaker and reached for a pair of glasses with large frames. As she picked up a box of oatmeal, the telephone rang.

"Hello?"

"Sally? This is Bev. What 'cha doin'?"

"Just making breakfast. How're you?"

"Fine. Making something along the line of a French omelet?"

"No, since Bill's first heart attack, oatmeal is standard fare. You know, heart attacks can be good for housewives," she chuckled. "Makes for simple breakfasts."

"Well, are we still on for brunch?"

"Wouldn't miss it for the world. I'll be ready in an hour or so."

"Great. See ya then." She hung up the phone.

She pulled a measuring cup of water from a microwave oven. No sooner had she stirred the instant oatmeal into the hot water than a large man about her age sat down at the kitchen table. Dressed in a white shirt and tie, he unfolded the morning paper and began reading.

She poured a cup of hot coffee into a ceramic cup and placed the bowl of oatmeal in front of him. After topping off her cup, she joined him.

"Aren't you having any breakfast this morning?" the balding man said, looking over his reading glasses.

"No. I'm not hungry." She smiled. "I'm tired of oatmeal, anyway."

He returned to the morning paper and began to read as he spooned the oatmeal. "What I wouldn't give for some bacon and eggs."

"Bev and I are going to the Outlet Mall this morning. Any special requests?"

"No, I don't need a thing." He thumbed through the classified ads in the paper. "Have you given any thought to another car? The old Cadillac has a lot of miles on it." Turning the paper toward her, he pointed at a car advertisement. "Do you like these new Impalas?"

"Oh, I really like the Cadillac. It's so comfortable, and it's the only car I feel safe in." She waved her right hand in a downward motion. "It's like an old friend. I'd really rather just keep it. Besides, the new ones are small and cost so much."

"Just concerned about it breaking down somewhere." He turned the page. "The thought of you standing along the interstate trying to get a ride home worries me."

"It hasn't failed us yet." She stood up and turned off the television. "Besides, I have my trusty cell phone."

His attention returned to the oatmeal.

She finished her coffee. "Guess I'll go put my face on. We're trying to beat the crowds."

"Okay. I'm going to work in a few minutes." His gaze never left the newspaper.

She walked down the hallway into the bathroom and began putting on her makeup.

A few minutes later, he called down the hallway, "Good-bye."

"Good-bye, dear."

The front door opened and closed with a familiar clunk. She heard his car start and pull down the driveway in front of their house.

Brushing her gray hair, she studied the mirror as if looking for a mistake in the makeup job. Then, she changed into a charcoal-colored dress hanging on a towel rack.

In the kitchen, she rummaged through the large black purse on the countertop, pulled a set of keys from it, and walked through the utility room into the dark garage. The door on the old blue Cadillac squeaked as she opened it and climbed in. The car rumbled to life, and she pressed the garage door opener clipped to the visor.

In minutes, she was out on the interstate traveling south toward the Outlet Mall.

She had only driven about eight miles when the engine died. "Oh great," she said to herself, "The one day I brag about the car."

Steering the car to a stop on the shoulder of the road required most of her strength. Now what? She looked down and fumbled in her purse for the cell phone.

A tractor trailer, loaded with forty-five tons of steel, traveling 100 feet per second, veered just far enough off the interstate to strike the Cadillac in the rear.

The car practically exploded. The tremendous energy catapulted it into a concrete bridge abutment thirty yards away. The sound of the crash was like a bomb going off. The front of the car struck the concrete, causing a second hard impact. The airbags deployed and powder from them created an artificial smoke inside the car.

Dazed by the shock but still conscious, Sally reached for the door handle, but it wouldn't open. The doors were hopelessly mangled into the rest of the car. To her horror, she discovered more than powder contributing to the smelly fog inside the car—there was real smoke.

Flames leapt from the underside of the Cadillac. Her eyes widened in horror and she beat furiously on the inside door panel of her car.

A man ran toward her car, stopped a few feet from the car, and held his hands in front of his face to shield himself from the incredible heat.

Sally screamed.

CHAPTER 7

The Automobile Manufacturers Casino

A car doesn't exist that could withstand an impact like the one described in the previous chapter. Bizarre accidents occur, but few could be more horrific.

The Automobile Industry has become increasingly aware of many safety issues over the years. They have significantly improved vehicle design and construction. In the last forty years, they introduced seat belts, shoulder harness, airbags, and cars designed to absorb impact. Most automakers crash test their products on a regular basis. Do you wonder why?

One reason is because they faced numerous lawsuits over safety issues. In view of the comments about our tort system in an earlier chapter, we do owe a thank you to trial attorneys for this increased attention toward safety. No doubt, legal action has resulted in positive changes in automobile crashworthiness.

It is unfortunate that the automakers must be sued in order to change safety standards because the litigation causes higher liability insurance premiums for the vehicle manufacturers as well as financial loss. These losses are passed on to the consumer through higher prices for their products. Once again, consumers pick up the tab.

Another factor affecting automakers and consumers is that insurance companies have realized that big differences exist in repair costs, based on automobile design. Insurers formed information exchanges with manufacturers for design improvement in an effort to improve safety and lower costs. In addition, crash tests conducted by the Insurance Institute and by the National Highway Traffic Safety Administration make that information available to the manufacturers and the public.

Yet, the costs aspect has not been fully utilized by the manufacturers. Can you guess why? It affects their profits negatively when repair costs are reduced.

The government sometimes controls automakers through regulations, but cars could be safer than they are. It takes more money to build safer cars, and the industry feels some customers are unwilling to pay more for cars.

For example, according to information on the Continental Teves Web site (www.contiteves-na.com), electronic stability control could save two thousand one hundred rollover fatalities a year. That number would probably be inaccurate today because in recent years more carmakers have built this technology into their automobiles.

The auto industry claims to be "consumer driven." They say consumers want SUVs, so the industry builds SUVs. The government enacted a "gas guzzler" tax on vehicles that are not fuel efficient. But, the manufacturers found a loophole in the law so that these type vehicles could be made and sold to us without paying the same tax as other gas guzzlers. That translated to a $1.1 billion loss of revenue that had to be made up by the typical taxpayer. Recently, demand for SUVs has decreased, so this revenue loss has also decreased.

Is the automobile industry truly consumer driven? Do the people who can afford these expensive cars control what the manufacturers produce, while passing on the costs to the majority? The auto industry, like most businesses, is profit driven. The average consumer is not served, but penalized. The use of the term "consumer driven" seems a feeble attempt to avoid responsibility.

The government requires passenger cars to meet federal standards for bumpers, but the SUVs are exempt. The cost to repair an SUV after a five-mile-per-hour impact is between four and six thousand dollars for most models (www.suv.org). Few people really need a four-wheel-drive, sport-utility vehicle, but these vehicles earn the manufacturers the most money. Why hasn't our government closed this loophole?

One reason we buy these cars is that we have allowed the ad agencies to pervade our thinking. Ironically, one theory is that most people drive SUVs out of fear. Many want the image created by ad men so that others will give them respect. After all, can't they do anything in the all-terrain, awesome, rugged, overcome-all SUV? Others are simply afraid of accidents and believe the SUV is safer. They certainly cost more to purchase, maintain, insure, and operate. Some models have been referred to as "OPEC Car of the Year."

Please don't misunderstand. SUVs are wonderful. In fact, who wouldn't love to drive an SUV to the great outdoors for recreational purposes or to have one for inclement weather? The point is that they should not be exempt from regulation, and all costs associated with their inefficiencies and dangers should be paid by those who own and drive them—not those who do not. However, when gasoline prices rise this problem is reduced because fewer people can afford to drive them.

Cars could be simply transportation devices for moving people from one place to another. In most cases, they don't really require a lot of speed or power. Yet many demand that these vehicles—especially sports cars—be loaded with both. Who needs a car that accelerates from zero to sixty in five seconds? What we should be looking for is a vehicle that can decelerate (stop) in the shortest distance.

What we, the consumers, really need are safe, energy efficient, durable, recyclable, comfortable cars. Yet there seem to be few produced. Rising fuel prices may change much of that.

Technology exists that could make cars much safer by enabling them to interact with the driver. Alarms could alert drivers who speed, follow too closely, or become drowsy. A vehicle could be equipped to refuse access to an impaired driver. Why not use this technology? The obvious answer is money but wouldn't a few lives be worth the price? Or are you one of those who would only say yes if you knew the life saved would be your own?

At any rate, the automobile industry will not do anything differently until we, their customers, insist. Auto accidents result in massive sales of parts by the manufacturers. Estimates of crash repair parts sales range from eight to eighteen billion dollars annually. It is difficult to assess the actual amount because these parts come from the original manufacturers, salvage yards, and aftermarket manufacturers.

One study figured the cost of new, original manufacturer (OEM) parts for a 2001 Chevrolet Cavalier to be $63,240, not including any labor or paint. The retail cost of the new car was $15,395 at the dealer. This is an example, and not used to single out this manufacturer, because results for most manufacturers are similar (www.ohioinsurance.org). There seems to be high profit margins in parts sales. If you worked in a profit-driven enterprise, such as an auto manufacturer, would you recommend making cars that were less expensive to repair?

Automobile accidents also serve to increase demand for new cars after the old ones have crashed. This generates business for them. Do not expect them to voluntarily reduce their profits.

American Highway Roulette generates big business for the automobile industry. Consumers, through the price of cars, taxes, and insurance premiums, pick up the tab. It seems readily apparent that our current system needs improvement and that will only happen through the efforts of consumers, voters, and taxpayers.

Motorcycles and ATVs are fun but offer very little protection and are extremely dangerous. Hazards that would pose no threat to a car can be disastrous to a motorcycle. Animals, loose gravel, slippery surfaces, and even crosswinds can cause a loss of control. And that's only a short list of potential threats to the motorcyclist.

Because of the smaller size of motorcycles, drivers of cars often do not see them. They can be obscured when traveling next to or behind another car. Many times fatal accidents have occurred to motorcyclists who practice safe driving habits.

If you must drive a motorcycle be advised that you should wear all the safety gear: helmet, eye protection, boots, leather pants, and jacket. They won't guarantee your life, but they will improve your chances for survival.

If you are a member of the automobile manufacturing community, I challenge you to build the safest car in America, regardless of vehicle class, and make it fuel efficient, environmentally friendly, durable, reliable, and economical to repair. This is what American consumers need. There is a good market for such a car, and those who have come closest to achieving these goals have been the most successful.

American Highway Roulette

CHAPTER 8

The old Volkswagen engine whirred as the young woman downshifted, using the gears to slow for a stop. The dazzling sunlight almost obscured the short line of traffic that was already stopped, waiting for the light to change. The white upholstery inside the faded red 'beetle-bug' had been soiled and split from years of use. It was hot inside the car because it had no air conditioner.

"Aunt Katie, why don't you have any child-ren?" A red-haired child sitting in the back seat asked, as the driver downshifted into second. The exhaust reverberated again in a reverse crescendo as the engine slowed the car. The windows were down and the worn exhaust had little muffling effect.

"Oh, honey, I haven't been blessed with any children yet." The petite driver of the VW strained to be heard over the noise as she shifted into first gear. "But, maybe one of these days..."

Another girl, sitting in the front seat, asked. "Do you want a girl or a little boy?"

"Oh, I'm not particular. I'd be happy with either one." Katie depressed the clutch again, and the car idled quietly in neutral. "As long as it's as cute as you two."

A large gray car was traveling fast from the opposite direction, approaching the intersection.

"Is Mommy going to come pick us up at your house?"

"Yes, as soon as she gets off work." Katie looked past the several cars ahead of her to the other side of the intersection. "I'm fixing dinner for us."

The speeding car hit a dip in the intersection. Sparks flew as the bumper struck the asphalt, causing the front of the car to go airborne. Surprised, the driver braked and turned the steering wheel. The car began to skid, rotating 180 degrees. The tires squealed from the sudden friction overload.

Kate watched, helpless. The rear of the skidding car struck the tiny Volkswagen with a loud boom. The force of the impact crushed the front of the VW and knocked the car backward. Everyone was thrown forward as if catapulted from a cannon. The steering wheel caught Katie's body, but her head struck the windshield with a sickening thud.

Both cars came to rest on the roadway.

A woman in a nurse's uniform jumped from a car that had been stopped at the light and ran toward them.

"Katie, I can't see! I can't see!" the bloody face of the girl in the twisted front seat yelled.

"Aaheee!" The sound came from the backseat.

Dazed, Katie glanced at the children and then struggled with the door until it opened. Standing, she spoke to the off-duty nurse who had just reached her car.

"It's okay. It's not really an accident," she said. "I just want to go home."

Katie collapsed, dead, her body making a hollow thump as it landed on the shoulder of the road.

CHAPTER 9

The Health Care Casino

No medical or health care provider could have saved Katie. Fortunately, they were able to help the two children.

The healthcare profession helps those injured in automobile accidents and also benefits from *American Highway Roulette*. The ambulance services, hospitals, doctors, radiologists, physical therapists, pharmacists, chiropractors (the list goes on and on) all make money from accidents. Be thankful that we have these people, buildings, and services in place. They save lives and make coping with injury more bearable.

Nevertheless, they are players in this game.

Let's look at an example of a problem in our current system. After an accident, a person who isn't injured but claims to be injured must be transported by ambulance to a hospital. The police do this as a matter of procedure because they have been sued when they didn't.

Once the "injured" party arrives at the hospital he or she must see a doctor. The doctor makes a diagnosis and recommends treatment.

If the injury is perceived as minor, the E/R doctor will refer the patient to his or her family doctor. Another set of x-rays is taken, and the family doctor gives his or her opinion, prescribes medicine, and/or physical therapy, and then sets a time for a follow-up visit.

Later, if the person still reports problems, the doctor may change medication or refer the person to a specialist, and/or physical therapy.

The specialist often orders expensive tests (CAT scans or MRIs) to see if the problems are more extensive than the simple tests indicated. In situations where simple tests show no severe injury, but the patient reports symptoms or pain where none would be expected, the doctor must continue to explore for other problems.

This is known as "covering one's behind." If they don't do something, they can find themselves in a lawsuit. The large number of lawsuits against medical providers results in exorbitant medical malpractice insurance. Medical providers pass this cost on to the consumer. So, not only do we pay for the cost of unnecessary treatment, we indirectly pay the medical provider's high malpractice insurance premiums.

Now, based on what we discussed about claim value in an earlier chapter, what would you recommend an injured party do? Yes, get as much medical treatment as possible. It might exponentially increase the value of the claim.

Let's say that the injured person decides to exaggerate an injury. At the scene, he or she tells the officer about pain in the neck area. A ride to the hospital, visit with the ER doctor, and x-rays will drive medical expense into the range of one thousand to twelve hundred dollars. Next, he or she will set up an appointment with his or her family doctor, who will also x-ray, prescribe medication, and physical therapy. If this person tells the doctor the treatment isn't helping, the physician will refer him or her to an orthopedic specialist.

Another alternative is to go to a chiropractor. I have seen chiropractic bills, submitted to an insurer that ranged from one hundred and fifty to six hundred and fifty dollars for the first visit. The patient is asked to return three or four times per week, and the costs run into the thousands for what is often considered a soft tissue injury.

Either way, it won't take long before this person's medical bills are four or five thousand dollars. Can you see some injustice and waste of resources within our system? In addition to the built-in incentive for greed on the part of victims, the same incentive exists for medical providers. If the patient has insurance and he (or his attorney) wants treatment, why not?

Then, there is the problem of an unscrupulous medical provider. According to information on the Coalition Against Insurance Fraud Web site, which quotes the Journal of the American Medical Association, nearly one-third of doctors exaggerate severity of a patient's illness to help the patient avoid early discharge from a hospital. The same source advises that nearly one-third of physicians say it's necessary to game the health care system to provide high quality medical care (www.insurancefraud.org). Consider that this points to another possible problem: unfair practices by insurers or other medical-payment providers.

Still, the National Health Care Anti-Fraud Association Web site (www.nhcaa.org), states that fraud amounts to 10 percent of U.S. health care expenditures. According to the Health Insurance Association of America, 80 percent of health care fraud is by medical providers, 10 percent by consumers, and the balance by other sources.

Many health care providers point to the legal profession and the insurance industry as the culprits. Yet, they will not take responsibility for the policing of their own industry. This problem goes far beyond the narrow scope of automobile accidents. It will take an informed public, motivated by the direct and indirect costs of health care, to make a difference in the interrelated professions of law, insurance, government, and health care.

A key problem with our health care system is that it is so expensive. We spend more than any other country, yet we get less value than most. In 2000 the World Health Organization within the United Nations compiled a report comparing the health systems of different countries. We pay the most and get much less in return.

One reason is the extensive lobbying by the health care industry. According to Congressman Carroll, during the 1998 through 2007 time period, the pharmaceutical industry spent $1,316,714,703; the hospital/nursing home industry $563,926,474; and the health professionals $531,096,203.

Why did they spend such large sums of money? To protect their business interests. Some may list other motives, but it seems apparent the reason has to do with the almighty dollar.

One reason for our expensive health care system is doctors want to earn a lot of money. Hardly anyone discusses this aspect, but it is an issue that needs attention. Is there a limit to what doctors should charge? Physicians in this country earn much more than those in other countries.

Certainly, if one has a particular talent and the market will support a higher price, that physician should be able to earn more. But we, the consumers, need to have access to doctors who can provide affordable health care.

Have you considered why the medical schools offer such limited enrollment? Surely we could admit more students to these schools and expand our training programs. Many qualified students are turned away each year. A greater supply of physicians and nurses could be instrumental in reducing our health care costs.

It has reached the point where American consumers are disgusted by paying too much for medicine, health care, and insurance. If professionals in these areas do not take innovative action to solve problems within their fields, consumers, through our government, will insist on legislative correction. Evidence of this phenomenon is reflected in the political platforms of both parties.

In general, the health care profession has demonstrated little concern about traffic accidents. Certainly, those people who witness firsthand the injury caused by accidents have empathy for those involved.

Still, *American Highway Roulette* is a huge source of revenue for the medical profession. Since the medical community shows little interest in change, shouldn't we, the people paying the bills, tackle these problems? What are our leaders in the medical field doing about it?

The American consumer needs an efficient and effective health care system available for all.

If you are a member of the medical community and wish to protect the status quo, I challenge you to formulate a workable plan to make health care affordable and available to American consumers, and I ask you to support efforts to reduce the numbers of accidents as well as the consequences.

The solutions are complex and are discussed in chapter 12.

American Highway Roulette

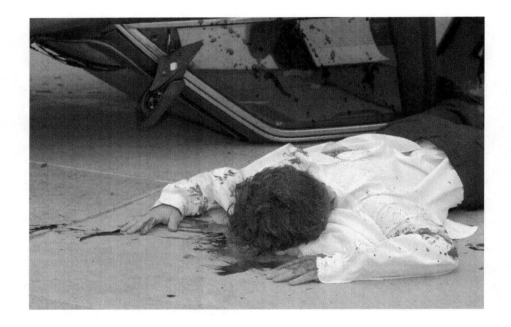

CHAPTER 10

The instrument-panel lights created an eerie glow on Dustin's smooth, pale, face, as he steered the Honda into the tunnel made by the car's headlights piercing the surrounding darkness. Gray asphalt slipped beneath the hood of the car. His veiled eyes peered from beneath blond eyebrows, alternating between his passenger and the road.

Beside him, Jason rocked in rhythm to the powerful sounds emanating from the custom speakers. His dark hair swayed in sync with the music as his head moved with the beat.

"Dude," Dustin said.

Jason paused. "What?"

"Are you sure you know where we're going?" Dustin pressed the eject button on the CD player.

"Yeah. Just a few more miles."

"This better be a good party. Hand me those CDs."

Jason pulled a container of CDs from the visor and sorted through them.

Dustin selected one and began to slide it in the slot. When he looked back to the highway, his headlights were pointing toward a ditch. He slammed on the brakes, causing a short shriek, as they left the roadway. The car became airborne for a second, the headlights shining into the night before providing a glimpse of the ground rushing toward them. Both yelled. *"Ahhhhhhhhh..."*

CCRRUUNNCCHH.

· · ·

While on patrol at four A.M., State Trooper James Stevenson caught a glimpse of a short set of skid marks leading off the highway into the grass—telltale signs of a one-vehicle accident. He turned the cruiser around and returned to the black marks. The shoulder was too narrow and steep to park, so he drove thirty yards beyond and parked on a gravel road.

Grasping his police issue flashlight, hat, and coat, he stepped into the cool darkness. He left the cruiser running with the radio turned up, the sparse chatter crackling noisily into the predawn silence.

He walked down the roadway until he found the tread lines on the shoulder. Taking very short steps, he climbed down the embankment, following his light. The beam revealed a smashed car lying upside down in the ravine. Training the flashlight on the car, he moved toward an unusual shadow on the ground. When the light crossed the face of a corpse, it created an image that he would never forget.

· · ·

(Note to reader—this is the last description of an accident that actually happened. You have read five of these and may be feeling like there were too many included. If so, please bear in mind that there were over 40,000 such instances in 2007, the most recent year for which statistics were available when this passage was written.)

CHAPTER 11

The Government Casino

The police pick up the pieces after automobile accidents. They see firsthand the aftermath of car crashes. Overworked and underpaid, there are too few to completely enforce the law. They do not have time to write tickets for following too closely, aggressive driving, and other such infractions until an accident takes place.

Budgets are big topics of discussion for all governments. Nobody wants to pay more in taxes to increase funding for hiring what might be perceived as a surplus of police personnel.

The effect is that traffic safety does not become a concern until some prominent member of the community is killed in an accident. Then, perhaps, a stop sign or road improvement might take place.

There are several things our government could do to decrease automobile accidents. Prevention is always the best solution. Most states have law that make speeding, following too closely, reckless driving, driving while intoxicated, and other causes of accidents illegal. After an accident, the persons involved may be given a ticket if it appears they violated these statutes.

But, these rules must be enforced *before* an accident, as opposed to after the fact. This is reactive law enforcement and we need *proactive* law enforcement.

Years ago, it would have been difficult for a police officer to prove one car was following another too closely. With today's inexpensive video equipment, it would be a simple matter. Police departments need more staff in order to enforce these types of traffic violations. The revenue they could generate might more than offset the cost.

One way to maximize enforcement without adding a lot of people is through *automated enforcement*. One method utilizes video cameras activated near stoplights at intersections. Others can record speed violations.

A study in Oxnard, California showed that red-light violations dropped 42 percent after cameras were introduced at nine intersections.

Some people feel automated enforcement violates their rights. But, when it comes to the lives and well-being of my family, I don't care whether another gives up the right to violate the law and not get caught.

Aggressive driving is a recent addition to the list of contributing factors to an accident. According to NHTSA, 66 percent of all traffic fatalities are caused by aggressive driving behaviors. Personally, I resent aggressive drivers and would like to see them ticketed. Someone driving seventy-five miles an hour or more, about ten feet from your bumper, could cause some serious problems if an unexpected event occurs. These drivers have no right to expose anyone to the risk of injury or death.

If enforcement measures were taken, traffic accidents would decline. Once people receive tickets for following too closely—before their accident—they will start allowing more space between cars. The rear-end collision is the most frequent type wreck and could easily be prevented.

Reducing the amount of traffic could also help reduce the numbers of accidents. If employers could allow flexibility in work schedules, perhaps "rush" hour could be minimized.

Similarly, during times of inclement weather such as snow and ice on roadways, work alternatives should be developed. Our weather forecasters generally provide reliable information, yet many employers take an insensitive position with regard to the danger present to human life and property. It is no secret that accidents increase dramatically during these hazardous conditions. Consumers unintentionally pay the costs of these accidents. Why should employers benefit? Legislation could be passed to change this practice.

More public transportation could also reduce the number of cars on the road. While most people think of the United States as a place without good mass transit systems, we actually do have a good one in place. The public school bus system covers almost every road in my area. Of course, how to expand on it and make it economically viable may be too much of a challenge for our community, business, and government leaders.

And, why do most school buses fail to provide seat belts? You could be ticketed for failing to buckle up your child, so why aren't buses equipped with this basic safety device?

Inconsistencies in law are a problem across the board. One set of rules would greatly simplify our legal system. Our current system of government dates back to a time when transportation and communication were problematic. In this modern age of technology, information can be transmitted quickly and efficiently. So, why doesn't the government take action?

Because politicians like the status quo. Local government jobs, county government jobs, and state government jobs employ more politicians. It is an inefficient system.

The 9/11 Commission pointed out the inefficiencies of one aspect of our government. The entire system deserves educated review and debate.

Our government has become the tool of popular opinion. Decisions based on public opinion polls often ignore the needs of the consumer. But no savvy politician will make statements contrary to public opinion. When constituents begin to get tickets for traffic violations that result in fines and possible insurance premium increases, politicians will feel pressured to change law and enforcement procedures. Political leaders need the determination to do what is best for all, not an aggressive few. Until we insist on making intelligent changes, it is unlikely that our government will take action.

Reducing speed limits could also reduce the number and severity of accidents. Change the speed limits back to fifty-five on the highways, enforce them, and then watch the fatalities drop. We'd also get the added benefit of better fuel economy, thus reducing our dependence on foreign oil. In some sparsely populated areas, it may not be necessary to decrease speed limits—only enforcement of existing law. Reducing speed limits interferes with the self-interest of many, so making this change will be difficult. Isn't the additional time spent driving at a lower rate of speed worth the lives, injuries, and economic costs of a higher limit?

Why don't we hear governmental discussion about how to make the highways safer? Do our politicians wish to maintain the status quo in order to protect their own interests? Without campaign contributions they can't be reelected. Many of the professions that profit are well organized to protect their own interests. Could it be that those in positions of power in our society have no real incentive to make changes that would affect the status quo? Or, are they, like many of us, simply ignoring the reality?

Thousands are killed each year. Our government has a traffic safety agency that has been quoted several times. Its job appears to be making us believe our highways are safe. But, our highways are dangerous, and you should believe that the events described in earlier chapters could happen to you or a loved one.

A few years ago, President George W. Bush spoke about another consumer concern—Social Security. This president was one of our highest-paid and most powerful government leaders, in a time when his party controlled Congress. He told you that your government could not get half the return on your investment that public sources could. Why not? If not, we need to elect new leaders. We do not have to accept poor performance from our government officials. They work for us.

This example above is intended simply to show that many, even politicians in both parties, accept an attitude toward government that is too tolerant of inefficiencies and ineffectiveness.

If you are a member of our government, I challenge you to educate yourself on the topics listed in this book and develop legislation that will reduce the numbers of accidents and minimize the consequences of those accidents that will inevitably occur. The American consumer wants an effective and efficient government. This applies to each and every program and includes people of all political ideologies. Those whose self-interests are threatened will oppose change, but we need leadership to serve the needs of American consumers.

The next chapter lists several things that can be done about automobile accidents. Be sure your elected representatives know what you want done about *American Highway Roulette*.

The Consumer Casino

According to some estimates, there are 246 million licensed drivers on our highways. In 2007, more than six million motor vehicle accidents occurred, causing almost two-and-a-half million injuries.

The National Highway Traffic Safety Administration estimated the economic cost at over $150 billion in 1993. By 2004, the number quoted by NHTSA and others is $230 billion. That means auto accidents cost American consumers over $7,293 per second.

The National Safety Council estimated the average economic cost per accident as $7,500 for no injury, $52,900 for injury, and $1,150,000 for fatality.

Clearly, automobile accidents cost American consumers a lot of money.

What can be done to reduce the numbers of accidents?

We know the causes of automobile accidents. Human behavior determines whether there will be an accident. Therefore the key question becomes how can we influence or change what people will do?

Religion has often been used to influence behavior. In the appendix, the Web site for the American Baptist Resolution on Highway Safety has been included. This document was created in 1966. There are also Resolutions for Highway Safety on the Southern Baptist Convention Web site. However, even such directives from one of the largest church organizations in our country have had little effect on drivers' behavior.

The following is the crux of the American Baptist Resolution: churches should stress the moral responsibility of every driver to develop safe and courteous driving habits; new drivers should be educated on the correlation of emotional disturbances to poor driving practices; legislation requiring safety standards for automobiles should be supported; and uniform laws and enforcement of penalties for traffic violations should be implemented.

This sounds simple and straightforward. It would reduce the number of accidents, if everyone adopted and acted in accordance with it. But, few Americans do and even many devout Baptists do not follow the resolution.

Education is another way to influence behavior. Almost every licensed driver has been exposed to driver's education courses. They are taught how to drive properly. Education helps but obviously will not ensure a change in human behavior.

In order to change specific driving behaviors, those who exhibit those behaviors must be punished and good driving behaviors reinforced. We have law that identifies those behaviors and provides penalties for their violation.

Therefore, the most effective action for prevention is simply enforcing existing laws regarding driving behaviors such as speeding, following too closely, reckless driving, driving while impaired, aggressive driving, and inattentive driving. These offenses should be punished more frequently and in the absence of an accident. Without proactive enforcement, these behaviors will continue.

Another very dangerous behavior is driving while drowsy. Very few states make this activity illegal. The National Sleep Foundation generates press releases that provide good information on this subject. In fact, they have set up a Web site dedicated entirely to this phenomenon (www.drowsy driving.org). Falling asleep at the wheel is an underreported cause of auto accidents.

The most common type accident is the rear-end collision caused by following too closely. The majority of these type accidents are relatively minor; yet do result in many of the whiplash type injuries. However, if you are rear-ended while making a left-hand turn, your car may get pushed into oncoming traffic, causing a severe accident. Once the following-too-closely law is enforced routinely, these behaviors will decrease, thus reducing the number of accidents.

For proper enforcement, video recorders must become standard equipment in police cars in order to document these infractions, so that tickets can be justified before traffic court judges.

Owning a device that assists a driver in circumventing the law such as a police radar detector should be illegal and those using them should be fined.

Law enforcement will require additional personnel, as well as automated devices, to aid in doing a more thorough job. In addition, signs should be erected to warn drivers of code enforcements in areas where strict enforcement guidelines will be followed.

How would we pay for these actions? It seems likely that the additional revenue from fines would offset the costs. If not, the reduction in accidents would certainly make extra enforcement worthwhile. It would make sense to conduct several pilot programs in some high-growth areas and study the results of such activities before implementing them nationwide.

One idea for an inexpensive and effective way to enforce driving law would be to encourage a citizen reporting system. Simply award a percentage of the fine for the violation as an incentive and many people would document driving infractions with their camcorders or cell phones.

To be sure, many will oppose increased enforcement because they view it as an infringement on their rights, freedom, and liberty. However, it has been suggested that one person's rights end where another's begin. Why should anyone have the right to place another at increased risk for property damage or injury?

Those in violation of sensible driving practices must be held accountable to protect those who do obey the law and practice good driving habits. Real change depends on having consequences for inappropriate behavior.

There are many ways previously identified that are aimed at reduction of accident causes. Installing devices on cars and highways that warn drivers when they begin to doze can combat falling asleep at the wheel. We have the technology to make a car inoperable by an impaired driver. It makes sense to incorporate all safety features possible in the manufacturing of our automobiles.

Send a message to manufacturers by your purchasing decisions. We may have to pay more for a car, but if lives are saved and injury minimized, why not?

Ask your elected representative to close the loophole on gas guzzlers and safety requirements. Vehicles should be constructed so that costs of repair for minor accidents will not be excessive. This would also help reduce the financial impact on the consumer.

Motorcycle use should be restricted to those over eighteen-years-old who have been educated to the dangers involved. A defensive driving course especially for operating a motorcycle is recommended as well as safety equipment. This is a very dangerous mode of transportation on American roadways at present.

Another way to reduce accidents is to restrict the use of cell phones. According to the Cellular Telecommunication Group, there are over 100 million wireless subscribers. Studies have shown that drivers using cell phones are more likely to have an accident. Perhaps one can think of circumstances where it is safe to talk on a cell phone and drive. But, most times it is not.

Texting poses an even greater risk. Some states have made this illegal, and the rest should follow their lead.

The one cause that gets a lot of attention is drinking and driving. Most states have increased fines and penalties with some success. Yet, some citizens contest the legality of sobriety checkpoints. If this method only saves a few lives, it is worth the time, loss of any legal right, and trouble.

Current law relies on the concept of punishment as a deterrent to inappropriate behavior. For the rational person, it seems to work. But, does the thought of punishment deter an addict? For some, yes. For many others, no. Many addicted people need help but do not want to change. Others may desire to lose their addiction but do not have the ability to change without professional assistance. Our current system is not very effective. Repeat offenders must be dealt with in a firm but therapeutic fashion. There are ongoing efforts to combat this problem through drug courts in some areas. Those with high success rates should be used as models. The sooner we restore these type people as functioning members of society, the sooner they can help support the system and end their patterns of abuse.

People in government can make a difference by improving our highways and making commonsense decisions on business zoning along busy roadways. Some of the poorly conceived intersections across the country are truly dumbfounding. Business leaders, city planners, and politicians in charge of this should be held accountable for their decisions.

Similarly, businesses could reduce the numbers of accidents by making concessions for attendance of employees during times when roadways are hazardous because of snow, ice, or other inclement conditions. We, the consumers, should not have to pay for the extra accidents that occur during bad weather simply so that businesses can profit. With current technology, many jobs could be done at home during such weather conditions.

Many safety efforts involve education. Certainly, we need to heighten awareness about the risks, especially seemingly benign dangers such as driving while drowsy. Real world consequences should be part of the education process. Seeing an auto accident victim makes a much more indelible impression than reading a book or listening to a lecture.

We watch movies where people perform fantastic feats behind the wheel of a car. Accidents don't usually claim the lives of movie heroes/heroines on film. And, since Hollywood tries to make films believable, what do we learn?

I can drive like that.

This contributes to the arrogance of "that only happens to someone else." I'm not suggesting censure, only stricter enforcement in order to remind people of the difference between fantasy and real world experience.

Another way to reduce the number of accidents is to incorporate graduated licensing for youthful drivers. This unpopular idea has been shown to have a positive influence, statistically. Those guilty of gross violations of traffic law must be held accountable at an early and impressionable age.

Similarly, retesting of senior drivers could reduce the number of accidents and save lives. Again, it will not be popular but should be done. The issue of freedom and rights will be raised, but how can one argue against the wisdom of such a measure? The Missouri model mentioned in an earlier chapter should be reviewed in a year or two for results. In view of the increasing numbers of older drivers, it is something we cannot afford to ignore.

Another important measure to reduce severe injury on highways involves *proper* regulation of the trucking industry. Too often, drivers are expected to spend an exhausting amount of time behind the wheel of a tractor and change his or her log to reflect breaks or down time. We need realistic, commonsense approaches in terms of driver qualification and trucking safety.

Reducing speed limits on heavily traveled highways can reduce the number of fatalities. It can also save gasoline, which reduces our dependence on foreign oil. Simple physics prove that severity increases when more energy is involved in an automobile accident.

In chapter 1, a trend in decreasing fatalities was noted. One of the possible explanations for this phenomenon is programs like the Click It or Ticket campaign initiated by police across the country.

What are some things that can be done to minimize the costs of American Highway Roulette to the consumer?

Several years ago, an innovative idea surfaced that was called a *pay at the pump* plan. Under this system, your insurance premium would be included in the price of a gallon of gasoline. Since taxes are already collected in this manner, the cost of collection would be minimized.

The beauty of the plan would be that everyone who drives would have to participate, thus ensuring that no driver is uninsured. A second major feature is that every person driving contributes to the system.

Some underwriting elements fit naturally within this idea. Larger, less fuel-efficient cars use more gasoline, cause more damage in an accident and the replacement cost of larger cars is usually higher. Those driving the most miles would pay more for protection. Those driving less would contribute less to the system.

This plan would save the auto insurance commissions earned by insurance agents and eliminate automobile underwriter jobs. It would also reduce and simplify claim representative jobs in that there would be fewer needed because there would never be a coverage question due to a person's failure to pay insurance premium.

The plan would also eliminate subrogation expenses, which result from efforts by insurers to recover amounts paid for damages caused by uninsured drivers.

Some would argue that the system would be unfair because an excellent driver with an excellent record would pay the same as a poor driver with a poor record. The argument is valid, but becomes clouded when one looks at the direct cost of uninsured motorist coverage combined with indirect costs currently included in health insurance and taxes. In other words, under the current system, even though as an excellent driver you pay less than an accident-prone driver, you still pay a premium for uninsured motorist coverage. And, some of your health insurance and tax dollars pay for damages and injuries caused by uninsured drivers.

This pay at the pump plan seems much more efficient. Since consumers are picking up the tab anyway, in one form or another, why don't we insist on efficiency?

There were several suggestions on how to handle claims under the pay at the pump system. One plan had insurance companies bidding for blocks of business. Others involved the government.

In order to provide uniformity, it makes sense to have this administered by the federal government, much the same as Medicare. This method would eliminate having two claim representatives from two different insurers handling the same claims. It would eliminate any bias on claim value due to a profit motive and also provide better uniformity in payouts.

We already have a federal agency, the National Highway Traffic Safety Administration, in place with good ideas about how to combat many of the problems described earlier.

Another great cost-saving idea is a true no-fault insurance system. Such a system would eliminate liability claims. The money paid into the system would be paid out in benefits.

Unfortunately, most no-fault plans in place today are inefficient and ineffective because they have been watered down to the point of losing the original benefit. Some serve in a counterproductive manner.

For example, consider the situation in Colorado. This no-fault plan provided $100,000 coverage for medical expenses caused by an auto accident. The coverage period for treatment of accident injuries was ten years. The plan was also very liberal, paying for such things as hot tubs, religious healers, health club memberships, as well as alternative and experimental treatments.

The important feature of this plan was the $2,500 threshold required before one could bring legal action on liability claims. Once the injured person accumulated $2,500 in medical bills, he or she could proceed to sue the other party. In other words, the fault component was never effectively removed from the no-fault law.

In today's expensive medical environment, it doesn't require much, if any, injury to accumulate $2,500 in treatment costs, especially under such a permissive plan. The people who exaggerate injury claims easily cross the threshold from no-fault to liability claims in this situation.

The result of this no-fault law was that it simply financed many, more expensive, liability claims. Now that you know how the claims "game" works, can you see how such a plan was a boon to the lawyers in Colorado and elsewhere? (If you had an accident in Colorado, you could be subject to the plan, even if you lived on the East Coast)

The average no-fault (they called it PIP - Personal Injury Protection) claim there was $7,749, and since 1997 the insurance premiums increased by 53 percent. It seems clear that in order for no-fault to work, one must take fault out of the system. In fact, the law in Colorado changed in recent years and costs are trending back to previous levels.

Equally clear, most Americans are reluctant to give up avenues for addressing grievances. A strong sense of justice has always been an important value to us. Therefore, an arbitration process or other fair method of dispute resolution should be incorporated so that injured parties have a way to resolve disputes with claim handlers.

In a true no-fault system, all liability disputes would be eliminated. What would be the point, if fault made no difference? Improper or intentionally malicious driving conduct would continue to be punished through the criminal justice part of our legal system.

Also, damage disputes could be minimized by including specifics within no-fault statutes about what is to be paid in benefits. Worker's compensation law does this to a large extent and could be used as an example for designing a new plan. Payment parameters specified by law are difficult to argue. There would need to be discussion and agreement on how to write in these specifics.

Fairness and equity require answers to the tough questions about claim values. Our political, legal, health care, insurance, business, consumer, and ethical leadership should be involved in this decision-making process. They owe this to us because we are paying for their services and we deserve fair, equitable treatment.

In its current state, the legal system is too inefficient as a vehicle for change. Our legal system needs an overhaul. One set of rules should be enough. Plus, we need to take the incentive out of frivolous litigation.

Perhaps the traditional one-third contingency fee should be modified to reflect the efficiencies that modern technology has provided to the legal profession. Currently, there are too many dollars going to attorneys and too few available for victims.

According to the American Tort Reform Association, our tort system returns less than twenty-two cents on the dollar for actual economic loss. On their Web site, one can see how tort reform benefits consumers.

In fairness, one can get opposing information from the American Association for Justice Web site (www.justice.org).

Another good Web site for information on this topic is Legal Ethics and Reform (www.legalethicsandreform.com).

Our current legal (tort) system has been coined a "crap shoot" in which anybody with enough money to file a lawsuit—in some cases only fifty dollars—can walk to the table (courthouse), roll the dice (sue some entity), and win big. The other side of this "crap shoot" must pay for defense, as well as risk losing the case.

A good way to change this and reduce costs in regard to auto accidents (and would help in other areas such as malpractice cases) is to write into law a statute that would require the losers of lawsuits to pay all court costs and attorney fees. This assessment should be shared by the losing entities involved and by their attorneys. That would have the effect of drastically reducing frivolous lawsuits. With attorneys having a direct stake in the outcome, both defense and plaintiff counsel would take a closer look at the merits of their cases before filing or defending a lawsuit.

The argument against this approach would be that insurance companies have shared resources to pay for those expenses. They have no personal stake in the loss since the cost is passed on to their policyholders. This advantage could be equalized by assessing those costs to insurance company employees making the decision to defend or sue.

In a similar vein, the "runaway" or "giveaway" jury syndrome must be addressed. Perhaps juries could be selected from a knowledgeable, responsible, reasonable, and a trained pool of people.

Of course, both problems would be eliminated by a nationwide, true no-fault system.

Presently, in our insurance business, we pay premiums for medical insurance benefits under our: auto policy, home owner's policy, health insurance policy, and worker's compensation policies. In addition, some of our tax dollars go to pay treatment costs for the aged, disabled, and indigent. Do we need all this overlapping coverage? Wouldn't it make sense to take a broader view on medical insurance issues as well?

As it is, we are paying for those who are uninsured in an indirect manner. Emergency medical providers can rarely refuse treatment to those without insurance. In many instances, they have to write off the treatment as a loss. That drives up the cost to the insurance-paying public. Health care providers can't do business for long without making a profit. We either pay directly or indirectly, so why not take action to manage health care? If health care were provided to everyone, perhaps preventative treatment would save money for treatment of some illnesses. If we have a healthier workforce wouldn't our productivity, and therefore our tax base, be greater?

Why is there resistance to a national health care system? Because it cuts into the profits of the insurers.

In our current system the consumer loses. For example, my wife and I are insured. When we have a medical issue, we go to a medical provider. They insist we sign an agreement to pay before treatment. We receive treatment, and our insurer is billed. The insurer almost always makes an incomplete payment. We pay a lot for insurance but usually end up paying a great deal for medical treatment not covered by our insurer. If we refuse payment to the medical provider, we could get turned over to a collection agency and have our credit ruined. The amount we are billed in any single instance isn't worth suing an insurer. It is doubtful any attorney would even take such a small case. It would translate into a lot of work for a little money. So we pay, and we are irritated at our loss of power. We are not alone and that's why health care was such an important issue in the last election.

Many argue that the government is inefficient and that the private sector can better handle health care problems. However, in my experience, the government has done a much better job. Medicare routinely reduces bills from hospitals, doctors, and other health care providers that private insurers simply pay.

This happens because Medicare has the legal power to reduce treatment costs and penalize those who break the rules. Insurance companies try to reduce inflated bills with medical providers, and some do but not to the extent of Medicare. Others attempt to reduce inflated bills but end up in expensive litigation if their policyholders are left with a huge amount to pay.

Most health care professionals are very aware of Medicare rules and regulations, and follow them for fear of being in violation. Medicare isn't perfect, but it is much more efficient and effective than the private sector.

The problem Medicare faces is not one of efficiency or effectiveness. The Medicare patients are all over age sixty-five, a time when medical problems become both common and severe.

Consider what is happening within the overall context of health care insurance. Private insurers selectively cover individuals until they reach the age where Medicare takes over. These insurers want to insure the young and healthy population. They take advantage of the profitable business but do not want to provide protection for those who need it most. That responsibility has been shifted to the government because too many seniors couldn't afford private insurance.

We've seen recently (AIG bailout) that insurers do not want to share profits, only the losses. The result has been that insurers can use years of profits to buy expensive buildings and pay high salaries while protecting their interests through lobbying and political contributions. The result has been a government that is not serving consumers by handling the problem of health care in the most efficient and effective way.

Similarly, fraud is a big problem identified in the area of health care and insurance. Insurers do not have the legal clout to effectively combat fraud. They currently rely on understaffed law enforcement agencies to pursue fraud. The fight against fraud will depend on development of investigative capabilities and effective prosecution by law enforcement. An integrated fraud unit, as part of our new Homeland Security, seems like a wise idea in terms of shared use of information resources.

At present, some people receive undeserved benefits. How can we minimize the incentive to freeload? One company utilizes a *work hardening* program. In some cases, injured employees report to a physical therapy clinic and receive a specialized in-house treatment designed to get them back on the job or to retrain them for different jobs. Since those people must report for therapy or retraining each day, any incentive for someone to make a false claim is reduced. Similarly, perhaps auto accident benefits, unemployment benefits, disability benefits, and others could be contingent on, or at least include, some aspect of work hardening or retraining for a different set of skills.

Most agree that we should not support someone looking for the "free ride." Everyone should be able to contribute something. And, if a person attempts to be productive, wouldn't we want to support them? Wouldn't you like support if you found yourself or your family in that situation?

One solution includes a combination of a private *and* public health care system. Public plans should include reasonable basic medical services, especially preventive, for every person. Public plans should have no costs, such as agent's premiums, advertising, or incentives (like conventions) included. They could even provide a provision not to sue except in intentional incidents.

A private system could be offered that would cater to those who want the right to choose exclusive doctors, hospitals, experimental procedures, extra benefits, etc. Then, each person can make the decision about whether they want to pay for the extra coverage. Private plans could continue to include current costs like advertising, lobbying, and incentives as mentioned above.

Under the current system, we pay health care premiums used by some for services that are unnecessary. Wouldn't you be perfectly happy to use doctors or hospitals that were willing to work for a reasonable fee, as long as the protection needed could be acquired?

If they made a mistake, would you sue them, if your family could get compensation for actual loss? Would you be willing to give up many of those "rights" that most of us never see personally such as getting rich from someone's mistake? This is a right that most people never use. If the right to sue could be removed from the public health plan, we could eliminate much unnecessary treatment and payment of malpractice premiums as mentioned in chapter 9.

Another problem is the high cost of health care. Some health care provider salaries seem very high. The prices charged for medications, supplies, etc. are exorbitant. But, since that is such a complex and large problem, it is beyond the purview of this book.

You can find opposing viewpoints on several of the conclusions in this chapter by visiting the National Motorist Association Web site (www.motorists.org). Your common sense should enable you to determine the truth.

This country is filled with brilliant minds in the fields of business, law, insurance, government, social work, and medicine. Why not establish a think tank composed of several representatives in each profession, as well as consumer groups, to address these problems and formulate intelligent solutions? Doesn't this seem like a good first step?

The purpose of this book is to heighten your awareness and give you a perspective on what happens every day. Ultimately, we are individually responsible for our own education. As a society, we have swept the whole phenomenon "under the rug."

For example, salvage yards are required to build fences that block the view of cars that are literally deformed. Accidents receive little media attention. Our heads seem stuck in the figurative sand, like ostriches.

Addicted to our automobiles and whimsical lifestyles, we ignore the truth, much like the gambling addict. The amazing thing is that we take risks for such minor reasons. A pack of cigarettes. A Sunday drive. We rationalize our actions for what we want.

Therefore, unless we act as alert human beings and work together to push our leaders into action, *American Highway Roulette* will continue in its present form. Think about it. Act on it. You can join some of the consumer safety groups on line.

Who are the losers in *American Highway Roulette*? There are the direct losers and the indirect losers. Those who have a loved one who has been seriously hurt or fatally injured are direct losers. The rest of us only lose indirectly, in a financial manner.

That brings us to consider the consequences of decisions on issues mentioned here. Some doctors, lawyers, auto company executives, insurance agents and managers, and legislators might lose money. Those in favor of keeping the status quo might argue that many of those might lose their jobs if our system changes. That could be true, but there should be new job opportunities as similar functions are handled by a government entity.

Most people who work in government agree that those who have invested in higher education and developed their professional expertise should be compensated in an above-average fashion.

Another argument used by those in favor of keeping the status quo is that these ideas reek of socialism. This word has acquired negative connotations. The proposals offered here could be better termed *consumerism*. Let our government do what is best for the American taxpaying consumer.

Often, you hear proponents of the status quo argue that the private sector is more efficient because of competition between insurers. Certainly, insurers compete with each other and that competition provides incentive to keep the prices they charge for their products as low as possible. However, the nature of that competition compels them to spend money on lobbying, advertising, and other marketing activities.

Some companies in this competitive environment fail for various reasons. When that happens, the claims of their policyholders are assigned by state pools to other insurers based on the amount of business they do in the state. Those costs are then passed on to the policyholders of the surviving insurance companies—the consumers. When all factors are considered, the private sector may not be more efficient. To do a mathematical calculation, better information would be required.

To further complicate the issue of private versus public health care system, organizations have been formed whose sole purpose is advancing political agendas. These organizations pay for misleading but effective advertising designed to influence thought. Because most of these organizations are funded by people and industries with money, there is a lot of propaganda circulating in our media. By contrast, there are few organizations with enough resources that advocate a public plan to present their case effectively to American consumers.

Another aged, private-sector argument holds that those who work in the private sector have more motivation. But, good management, whether in the public or private sector, should be able to motivate employees. If auto insurance depended on sales ability, perhaps the argument would carry more validity. However, since auto insurance has become a requirement, what role does a motivated salesman play? And, if the private sector can do the job more efficiently, why are they so afraid to let the government compete with them?

The profit motive drives business in the private sector. Businesses can't survive without profit. But, it should not be a motivating factor for an employee who is paying auto insurance claims. Unfortunately, this does happen.

In this author's opinion, effective government could get the job done less expensively. If not, the executives of our government should take proactive measures, such as firing ineffective administrators and employees, as happens in the private sector.

Isn't our government charged with regulating and protecting our society? Does democracy include the right to drive with disregard for the safety of others, without interference from the government or any other entity? A democracy should be the ideal political system for meeting the needs and wants of consumers. Unfortunately, our government seems heavily influenced by special interests.

It works like this. To get elected, a candidate requires a good deal of money to buy advertising in our media. Large corporations and special interest groups contribute those funds. Therefore, it is difficult for politicians to act in the best interest of consumers.

So the question becomes how much, as a consumer who currently picks up the tab, are you willing to pay for subsidy of the current system? As mentioned earlier, you pay multiple insurance premiums and taxes. Add them up. Are you getting your money's worth?

If prevention measures would be implemented to change the driving behaviors that cause automobile accidents, it is likely we could reduce the numbers of accidents by at least 30 percent. If so, twelve thousand lives could be saved each year. One of those could be yours. One hundred and seventy-five thousand injuries might be prevented. Close to two million accidents might be eliminated. That would save a lot of money paid out in claims and make a dramatic cost reduction to our economy.

In addition, those who choose to behave in harmful ways would contribute to our local governments when they are ticketed for their dangerous acts.

We will not be able to totally eliminate automobile accidents. Human beings will make mistakes. But, those mistakes can be minimized if the measures to reduce the consequences are implemented. More money could be placed in the hands of the victims. This savings could make another 30 percent difference in what you pay for protection.

Combining all the recommended prevention and minimization measures could result in lowering your cost by 30 to 50 percent, in my opinion.

These are tough problems and require thoughtful solutions. Many large organizations want to protect their industry's financial interests. They are not going to change voluntarily.

Since the consumer picks up the tab, there is no reason we shouldn't insist on minimizing this phenomenon as much as possible. Everyone will not agree on the methods of changing the status quo. Yet, you can see that there is need for change. Let's demand our money's worth. Seek good information, give it serious deliberation, and discuss with your friends. Then, write or phone your elected representatives and voice your opinions.

You have been exposed to the costs and consequences of our addiction to a deadly game of chance. We're not giving it up. But we can minimize the terrible results of automobile accidents and lower the risks. We must fight this deadly form of gambling and stop the killing, maiming, and injury on our highways.

If you are an American consumer, a person who wants efficient and effective insurance, law, health care, automobiles, and government for all, I challenge you to take action for change. Those whose self-interest is threatened will misinform and use hot-button emotional issues to sway you, so you'll have to seek the truth in a critical manner.

We can make a difference, but, as a fellow consumer, your help is needed.

American Highway Roulette

CHAPTER 13

Game Over

If you are so unfortunate to be someone who has lost a loved one to an automobile accident, your life will be changed in many ways. Your beliefs about the world will be challenged by questions, most of which begin with the simple word "why."

Why did it happen? Why did he or she have to die or become injured, crippled, or maimed? Why did this happen to me? It is difficult to answer questions like these or to find meaning in senseless tragedy. Typical answers do not hold up under logical scrutiny.

Most people believe in a higher power or Supreme Being. Often, their faith is shaken. For the purposes of discussion, the word God will be used to describe that entity, although you could interchange other names from different religions.

In many instances, people assert that God took their loved one. Why would a God, an all-powerful Creator, cause the death of a person, when almost all religions insist that God has told us not to kill?

God gave each spirit life, so why would God randomly eliminate more than forty thousand human beings a year over the past twenty years? Most of them hadn't lived long enough to realize their divine purpose. God may not appreciate being blamed for all this tragedy, pain, and suffering. God did not cause all these untimely accidents.

Some say that God took those spirits in order to punish those left behind. Surely, some might have deserved punishment but that would not make sense either. If God wanted to punish us, wouldn't the offender be punished directly instead of indirectly?

As an analogy, would it make sense for us, in our legal code, to punish a family member instead of the perpetrator? Of course not. The truth is that in the majority of traffic accident victims that I know, the families led exemplary lives on many levels and would be the last on any punishment list.

On the other hand, some theorize that God only takes the good. True, a vast number of those who have died were wonderful human beings. Yet, obviously, many great people have remained and lived long lives. And, of course, there were some exceptions, because some not-so-good humans have died in car wrecks, too. Clearly, this source of death in our culture doesn't exclusively take the good.

Others believe automobile accidents are a matter of destiny. We've all heard local sages sigh and tell us, "When it's your time to go, it's your time to go." These words have been used as the justification, or flawed logic, for doing something hazardous. This theory doesn't make sense, either.

If accidents were destined, why did the numbers of people killed in car crashes increase directly in proportion to the numbers of cars and drivers on the road? In other words, if destiny were at work, the pre-destined would die each year, regardless of the numbers of cars, drivers, and miles driven. Why did the number of fatalities decrease when the speed limits were reduced to 55 mph? Either fewer people were pre-destined to die in auto accidents that year or slower driving resulted in fewer deaths. Perhaps one could theorize that the law changed those persons' destinies, but that argument seems illogical. Similarly, the numbers of fatalities decreased in 2008 due to: fewer miles driven because of high gasoline prices and economic downturn; safer vehicles; and programs such as *Click It or Ticket*. It seems clear that less driving, slower driving, safer vehicles, and prevention efforts result in fewer and less severe accidents.

Perhaps some are destined, but certainly all are not.

Even in years preceding 2008 we have been able to influence fatality incidence by making rational choices about safety. Government statistics support the fact that there have been fewer auto accident fatalities, per million miles driven. And there is no doubt that improvements in auto safety design have made a difference.

If destiny were at work, wouldn't it operate independently of all the other things we do? Please reflect on this, and you'll realize the absurdity of such a theory for auto deaths. It is impossible to know whether some people may have been destined to die an early death. Still, our logic tells us fate does not explain auto accidents.

By and large, destiny is simply another theory to avoid personal responsibility. In other words, we do what we wish, without regard to consequences because, after all, when it's our time to go, it's our time to go.

The truth: we have the power to choose. These accidents don't have to happen. We make decisions, as a society, regarding automobile safety, highway design, law enforcement, education, and other measures affecting prevention.

Individually, we choose what we drive, how we drive, when we drive, and where we drive. These individual and collective decisions create a combination of events and situations that dictate the consequences of *American Highway Roulette*. Our desire to place the responsibility elsewhere reflects the immature nature of humanity.

Therefore, let's seek to mature, accept our responsibility, and take charge of our future.

We cannot change what has already happened. The loss in your life can be dealt with in a healthy manner through the grieving process. A counselor, therapist, or pastor can be helpful in coping with death. Support groups are very beneficial and available in most communities. You are not alone in your pain.

Perhaps you can provide meaning for your tragic loss or for the millions who have died in this way by taking action toward preventing these tragedies in the future. Do not allow yourself to be fooled by the illogical explanations of those who would have you ignore or accept the problem as one outside our control.

If you are lucky enough to be reading this without having lost a loved one to an auto accident, perhaps as you read the chapters describing actual accidents, you had a glimpse into the experience of others. Please bear in mind that just like you, those living human beings didn't expect to die in a car wreck. Keep in mind that the very people you might help or save may be those in your own circle of love.

Most of all, do not forget that *American Highway Roulette* is a random, indiscriminate, serial killer. Play wisely.

HOW TO IMPROVE OUR ODDS AND LEAVE LESS MONEY IN THE CASINOS

ACCIDENT REDUCTION
1) Enforce existing law:
   Ticket aggressive drivers
   Fine tailgaters
   Ticket speeders
   Penalize reckless driving
2) Increase police staffing
3) Use automated enforcement
4) Adopt the American Baptist Resolution
5) Use safety technology
6) Build safer cars
7) Restrict cell phone usage
8) Increase checkpoints for impaired drivers
9) Treat addicts
10) Hold leadership accountable
    Improve roadway and zoning design
    Only necessary people drive during inclement weather
11) Educate youthful and elderly drivers
12) Graduated licensing for youthful drivers
13) Test senior drivers based on need
14) Regulate trucking industry properly
15) Reduce speed limits
16) Fine those who use devices such as radar detectors to circumvent speed regulations.

REDUCTION OF ACCIDENT CONSEQUENCES
1) Pay at the pump plan
2) True no-fault legislation
3) Tort reform
4) Insurance reform
5) Fraud investigation
6) Work hardening
7) Health care reform
8) Legal reform

SAFE DRIVING TIPS

Always check intersections before entering them, even if you have a green light. People often run red lights. You may have the legal right-of-way, but not being at fault offers little consolation if you are killed or severely injured in an accident.

Always check for trains at railroad crossings. Do *not* depend on warning devices.

When driving on interstate highways, avoid traveling in groups of cars that may be tailgating. If a car tailgates you, increase your speed (within reason) until you can safely get out of the way.

Practice defensive driving. Never assume that a driver will turn just because he or she has a signal blinking. Imagine what could happen while you are driving at any given time and determine what evasive action you could take in that situation.

Never allow yourself to become angry while driving. You can choose how to react to any situation. In the case of an aggressive driver, why should you allow that person to change your emotional state? There is no good reason to do so. Anger will do nothing to serve you or change the behavior of another driver. The smart thing to do is remove yourself from the situation that may be dangerous and maintain a calm internal state.

Always signal your intention to turn, brake, or change lanes.

Avoid driving while tired, especially late at night. Reaction time may be diminished and risk for falling asleep increases.

Do not talk on a cell phone while driving. If you must communicate, find a safe place off the roadway, such as a parking lot, and make your call. Do not text message while driving.

Always look **twice** before entering a roadway. You rarely miss seeing an oncoming vehicle, but failure to notice a vehicle once may be your last opportunity.

Complete a vehicle inspection before driving. Check tire pressure and wear. Use a tire pressure gauge and tread wear gauge at least once per month. Measure tread wear on each side and in the middle. If one side wears faster than the other then schedule an alignment. If the front tires are wearing faster than the rear, then rotate your tires. Most dealers recommend rotation every 5,000 to 8,000 miles. Change tires when tread depth reaches minimum levels as recommended by your tire manufacturer. Good tread depth and proper inflation helps provide optimum handling and braking. This habit can improve safety and save money. Also check windshield wiper blades, windshield washer, positions of mirrors, turn signals, brake lights, headlights, and oil periodically. Keep your car well maintained. Be sure no frost, fog, snow, or ice obstructs your view in any direction.

Allow at least one full-sized car length for each ten miles per hour that you are driving. If you change lanes, be sure to leave plenty of space between your car and vehicles in the other lane. Allow extra distance for large vehicles such as semi-tractors or recreational vehicles. Although they have good braking systems, the extra weight can result in longer stopping distances.

Do not carry loose items in your vehicle. One man was fatally injured by a loose jack he failed to secure in the rear of his vehicle.

Do not drive if you are emotionally upset. If you have a disagreement with a passenger, find a safe place off the roadway, such as a parking lot, and discuss the problem there.

Do not listen to music at a high volume. This may impair your ability to hear warning devices such as sirens that could alert you to a dangerous situation.

Use caution when approaching a disabled vehicle. If possible, change lanes. In some states, the law requires you to do so.

Check your rearview mirror consistently to maintain awareness about what is going on behind you.

Avoid, if possible, driving on weekends during hours when your risk of encountering intoxicated drivers increases.

Check your gauges regularly as you drive.

Do not drink and drive. Similarly, don't use legal or illegal drugs that may impair your driving.

Do not overdrive your headlights. When driving at night the question to ask is: If an un-illuminated object with no reflectors lies ahead in the roadway, will I be able to see it in time to stop before colliding with that object?

Avoid driving during heavy rain or other inclement weather. If you must drive in those circumstances, do not use your cruise control. If your cruise control downshifts and increases engine power, you may lose traction with one or more tires, causing you to lose control. Also, avoid areas prone to flooding during heavy rain. Do not drive through water covering a roadway.

If driving long distances, stop regularly to avoid fatigue. Drinking water will help ensure you stay properly hydrated and cause you to stop periodically to use a restroom.

Ask your mechanic to check radiator fluid level, belts, and radiator hoses when servicing the car. Many recommend flushing the radiator and replacing fluid and hoses at least every four years. Some manufacturers recommend changing radiator coolant every year. It contains a lubricant that extends the life of your water pump. Also consider replacing your car battery every four years unless your battery warranty extends past that time. Radiator hose leaks or failures, water pump problems, and broken belts account for a large percentage of the problems that leave a motorist stranded on a long trip. The shoulder alongside an interstate highway is a very dangerous place. Similarly, by changing the battery before it goes dead you might avoid a problem. Think, Murphy's Law: anything that can happen will happen, at the worst possible time.

If you find yourself getting fatigued and unable to pull over, change the focus of your attention in short quick intervals. For example, alternate between looking at the road, instrument gauges, mirrors for split-second intervals. Also, allow more distance between your car and one in front of you. If you have difficulty holding your head up, keeping your eyes open, or if you find yourself drifting to the shoulder, get your car off the road as soon as safely possible. These are danger signals that must not be ignored. Do not drive again until you regain alertness.

If you are taking a trip, familiarize yourself with a map before you begin to drive. If you must read a map while on the road, pull off the road to a safe place. Driving while map reading can be hazardous.

Always wear seat belts with shoulder harness and insist that all passengers use them.

Operate lights during any time where visibility is not ideal. In some states the law requires use of headlights any time windshield wipers are needed. Even in sunny weather, lights increase the visibility of an automobile.

Allow extra stopping distance if the roads are wet. In hot weather be especially cautious after rain begins because high heat can cause oils to be drawn to the surface of some asphalt, creating a very slippery road surface.

Parking lots deserve special attention. Drive slowly and watch carefully. If you see someone backing out of a parking place, stop, and wait for that person. Do not assume that just because you are in the driving lane that you have the legal right-of-way. Both parties have an obligation under the law to keep a proper lookout and take any reasonable action to avoid an accident. Keep a sharp lookout for children and people in wheelchairs.

Of course, observe all traffic law while driving. Common sense dictates that speeding, driving too fast for conditions, following too close, etc. are hazardous activities that can also result in a fine. (See Ideas For Parents)

Read the owner's manual for your car and be familiar with all warning lights and safety precautions listed by the manufacturer. If driving a car not owned by you, get familiar with the features of that car and adjust all mirrors for optimum visibility.

IDEAS FOR PARENTS TO MANAGE APPROPRIATE DRIVING BEHAVIOR

Please note: experts have written many good books on behavior. Many professionals are available to assist parents with the establishment of behavior management with their children. The author does not claim to be an expert and makes no promises as to the effectiveness of methods listed below.

Probably the most important thing you can do as a parent is to set a good example or what many refer to as *modeling*. I clearly remember driving with my son when he was very small. He observed and commented on my driving behavior. He'd tell me if I was speeding or if I went through a yellow light—even before he started school.

If you tell your teen to do one thing, but you don't follow your own advice, what message does that send? That you don't really believe what you say, so why should he or she? Don't be a hypocrite.

Look for opportunities to demonstrate good driving practices. For example, if an aggressive driver cuts you off, say something intelligent like, "That behavior can make me aggravated, but I choose not to let another person control my thoughts or feelings."

When our children reach sixteen years of age, they can't be considered children or adults, so parenting during the teen years can be problematic. Many believe that an individual should be considered an adult at age eighteen. By then, most have been through an education curriculum and should be able to function successfully. However, it is very important for parents to have instilled desired behaviors in their young adults to enable them to behave in ways that will keep them healthy and self-fulfilling.

Most teens have led a relatively sheltered life and lack what some adults refer to as real world experience. They sometimes have attitudes that bad things will never happen to them. Another common feeling is one of invincibility, especially for young men. These ways of thinking and feeling sometimes result in taking foolish risks.

Some teens are also influenced by what they have seen in movies, television shows, and advertising. Often risky behavior such as speeding can be translated as cool, powerful, and a boost to self esteem. As an example, look to coming-of-age movies like the classic *American Graffitti*.

Recent studies have also indicated that brain function may not be fully developed in some teens until they are much older than sixteen. And many teens are not emotionally mature due to various reasons such as hormonal irregularity or imbalance.

Since automobile accidents are the leading cause of death for these young adults, the behaviors that contribute to accidents are the focus here. In order to engage your teen, it would be good to ask him or her which behaviors cause accidents. If you've read this book, I suspect you will be able to add to the list.

Most behavior can be managed through positive and negative reinforcement of those behaviors. However, you can be viewed as a manipulator if you choose to apply behavior modification techniques without being completely honest and open in an upfront manner with those with whom you have relationship. Therefore, you must explain why these behaviors are important to you, what impact misbehavior might have on both you and the teen, and develop a plan to achieve the goals you will have agreed upon.

Because most teens do not have jobs or other financial means, most of the financial consequences will fall on you, the parent. This is why you must address these behavior issues with your loved one. Even if he or she is eighteen, many will not have the means to shoulder all possible responsibility. In order to instill the correct behavior patterns, the teen should share the consequences, good and bad.

It is important to be very specific about consequences for each behavior. Some behaviors carry greater consequences than others. Since the teen cannot share the consequences of positive behaviors, like keeping insurance premiums lower, you must create positive consequences for him or her.

This is a good reason why you should try to learn what would motivate your teen. Each teen has different wants and needs, so try to encourage open and honest communication.

The next step is to write up an agreement outlining the specifics. Setting expectations are very important in any relationship, and especially with this age group.

Consider whether you want to meet after a certain time and revisit or renegotiate terms of the agreement. You may also need to brush up on current law in your state as it applies to teen drivers.

For example, many states now have graduated licensing mentioned earlier in this book. Some have limits on the number of underage passengers that can occupy a car while being driven by a sixteen-year-old driver. You need to know and convey any legal restrictions applicable to your teen driver.

Then, you need to physically spend time in the car observing driving behaviors and explaining undesired behaviors. Be sure these behaviors are understood. The more training received by your teen driver, the higher the probability for successful driving experience. Don't assume your teen will learn all that is needed from the driver's education course taught in most schools. As a parent it is your responsibility to take an active role in preparing your teen to practice safe driving habits.

As always, you can use recognition, approval, and affirmation as positive reinforcement. But, try to catch your teen doing something right and provide some concrete, specific reward that you have previously agreed on.

Try to schedule your personal driving activities, so that you can have an opportunity to see your teen's random driving behaviors when you are not in the car with her or him.

You will have to decide on a time frame for administering rewards. The shorter time frames are best as it keeps the teen aware of consequences. Certainly, when an insurance premium is due would be a good time for administering reinforcement.

Of course, any trigger event such as a speeding ticket should entail negative reinforcement.

You might think that this is all too formal or too structured or just plain unnecessary. You may feel that just discussing the matter will be enough. But, trust me, if something happens, there will be argument about exactly what was said.

If you have your agreement in writing, you will be likely to avoid argument. By comparison, think about all the behaviors listed in the mortgage agreement on your house. For most parents, their child's well-being is much more important.

Most people recognize the following behaviors as those that cause automobile accidents. These are listed on most police reports in the contributing factors section.

Inattention - This is a vague behavior description and should be discussed in more specific terms. It could include talking on a cell phone, being distracted by an object or event off the roadway, eating while driving, becoming absorbed by music or audio book, arguing with a passenger, putting on makeup or shaving while driving, having several friends in the car, as well as any other activity that takes one's attention away from the task of driving safely.

Following too closely - This is self-explanatory, specific, and receives much attention by driver's education professionals. One thing to mention is that when someone cuts in front of a driver, he or she must increase the distance for following the aggressive driver to compensate for the reduced space between cars. On the other hand, stress the importance of allowing plenty of room when passing a vehicle before pulling back in front of them. Often, side-view mirrors provide the illusion of adequate distance.

Failure to yield the right-of-way - This is also specific and easily understood. However, a good discussion of parking lots, and who has the right-of-way may be helpful. The fact that a person is in the driving lane does not necessarily mean he or she has the legal right-of-way. He or she may have a duty to stop if he or she sees a car backing out of a space. Also, a good discussion of the proper way to use a turn lane is recommended. Please note that some variation exists from one location to another. It is also worth repeating that just because one has the right-of-way, he or she should always check for other oncoming traffic. Emphasize courtesy to other drivers, especially those trying to maneuver an eighteen wheeler, bus, RV or other large vehicle through traffic.

Driving too fast for conditions - This covers several situations. Rain, ice, snow, sleet, fog, or smoke can create an environment where speed must be reduced for safety. The point to stress is that a driver must consider every driving situation and use common sense to maintain an environment that is safe for everyone. In some cases, that means making a decision not to drive. For example, if you live in a hilly area and snow creates slippery roads, and you only have a rear-wheel-drive vehicle, your chances of an accident dramatically increase. Ask your teen to consider asking a friend who may have an all-wheel or four-wheel drive vehicle for transportation if they must be somewhere.

Careless driving - This covers several situations where a driver acts with disregard for the safety of others. For example, driving forty mph on a street with this speed limit may be too fast if a crowd is moving on a sidewalk next to the street.

Aggressive driving - This includes behaviors such as tailgating, constantly changing lanes to advance in traffic, flashing lights at slower drivers, yelling or using obscene gestures, slamming on brakes, cutting off other drivers during lane restrictions, and other offensive driving behavior. These drivers exhibit impatience and are often under some sort of emotional duress.

Speeding - Driving above posted speed limits is dangerous because it results in greater stopping distances, increases the chance of losing control of a car, and reduces the opportunity to avoid unexpected hazards, such as animals or objects ordinarily unexpected on a roadway.

Some other specific behaviors commonly listed on police reports include: driving without lights, failure to dim headlights, disregarding a stop sign, disregarding a yield sign, disregarding a traffic signal, driving on the wrong side of the road, driving the wrong way in one-way traffic, improper turn, improper lane change, improper passing, prohibited U-turn, defective lights, defective brakes, improper backing, failure to signal, disregarding an officer or flagman, cutting in, impeding traffic, improper parking, and of course, alcohol and drugs. Make sure your teen understands each of these driving problems.

Naturally, you will explain that police officers will write tickets for those who they discover committing these infractions. Not only will a fine be involved, your or their insurance rates may be affected. This may also be a good time to remind your teen that if he or she allows another driver to use your (or their) car and that person has an accident, your or their rates may also be affected.

As mentioned earlier, one of the best things you can do is lead by example. Drive in a manner that you expect from them.

Important parenting decisions are sometimes influenced by thoughts and attitudes that are not well thought out. Some want their teen to begin driving as soon as possible in order to make it more convenient for the parent. Others select cars for their teens based on how cute the car might be, how economical, or if it lends to the family's status versus what might be a safer car. A few believe that their teen will automatically become a better driver as they get older. If a parent is uneasy or uncomfortable riding with their teen they should certainly work on driving skills before they allow the teen to drive alone or with friends.

The teen years can be emotionally difficult. Try to be aware of any issues and help your teen deal with them effectively. Professional counseling is a good idea for anybody and as a parent; you could be part of a problem. Take a proactive approach and you may increase the probability of your teen becoming a happy, healthy, and successful adult.

Below is one possible approach that you might modify for your use.

Step One: Sit down with your teen in a relaxed setting with no distractions to discuss driving behaviors. An opportune time might be when your teen wants to discuss getting a car or driving your car. Explain that as a parent you want him or her to behave certain ways while driving a car and that it is important enough to both you and them that you want them to participate in the rewards and consequences. Tell them that an auto accident is the greatest threat posed to their health and well-being.

Step Two: Encourage an open conversation about what your teen wants and needs, as well as what you want and need in terms of their behavior. Ask follow up questions to learn more about the nature of their wants and needs. Be open and noncritical at this point. Also encourage your teen to ask questions about the behaviors you want. Explore other possible rewards or privileges you might be willing to offer as well as several possible negative reinforcement that may obtain desired results.

Step Three: Decide what positive and negative reinforcement that you and your teen can realistically accept. Please note that severe negative reinforcement may only provide an incentive for your teen to conceal any undesired behavior from you. Discuss and agree on appropriate positive and negative reinforcements with your teen, if possible.

Step Four: Write up an agreement that you both can live with then ask your teen to sign it.

Step Five: Follow up periodically as specified in the agreement. Provide positive or negative reinforcement depending on the outcome for that time period. Determine if the agreement is effective and whether additional consideration or adjustment may be needed.

Even if you elect not to use this agreement, please consider how you can obtain the desired behaviors from your teen to optimize the outcome of his or her driving experience. This could be one of the most important things that you can do as a parent. It could be a matter of life or death for your loved one.

It could also be one of the most important things you can do as a responsible citizen for others who have family and loved ones on our highways.

CONSIDERATIONS IN BUYING INSURANCE

If you are concerned about the financial condition of an insurance company, you can check ratings by companies such as A.M. Best, Standard and Poor, and J.D. Powers. These organizations all have Web sites.

If you are concerned about customer service, you can check with your state insurance department, Better Business Bureau, the Internet, and anybody you know who has experienced a claim. You can also inquire about claim-handling procedures at an agent's office or visit an insurance company's Web site.

If you are concerned about price, shop around. Just be careful to compare coverage and the amount payable so that you have an accurate comparison.

For good in-depth information on purchasing insurance, please visit the Insurance Information Institute Web site (www.iii.org). It contains a great deal of useful information about insurance.

If you have little in the way of assets, you may not want to purchase more than the minimum required liability coverage. If you own assets, you'll want to evaluate the legal climate where you drive, the risk factors posed by your driving habits, the type car you drive, and the cost of extra coverage. Some people are more comfortable taking risks than others.

Before purchasing protection for your needs, review your financial situation with anyone who would be affected should you become injured or killed. Consider your income, the type work you perform, living expenses, debt, other responsibilities, savings or other resources, and how you might be able to deal with a severe injury. Because these factors change throughout life, you should review your needs periodically.

There is no magic formula for how much insurance you should buy. For most people cost and comfort level with risk determine what they choose. The important thing is to do this before an accident occurs.

One way to save money on insurance is to purchase coverage with high deductibles. Just be sure you have enough money available to cover the deductible should a loss occur.

COMMON INSURANCE TERMS

Betterment: Also known as depreciation. If replacement of a new part results in placing you in a better financial position, then you may have to pay a part of the cost of that part. For example, if you have a three-year-old battery, and it is damaged in a collision, you may have to pay a part of the replacement cost.

Capital Stock Company: These companies are set up like other corporations with stock issued, purchased, and sold in financial markets.

Coverage: The protection your insurer's policy agrees to provide.

Claimant: Any person who has a claim to benefits provided under an insurance policy. The most common reference is to a person who has a third-party claim based on liability.

Deductible: The part of the loss you agree to pay.

Exclusions: Situations in which the insurer will not pay insurance benefits.

Indemnity: Restoring a person to the financial position he or she enjoyed just before a loss. Insurance will indemnify you but is not intended to place you in a better financial position.

Insurance: Protection against loss. The losses of a few are shared by many.

Insurance policy: The contract that specifies the agreement between you and your insurance company.

Insured: any person who qualifies for benefits or protection in an insurance policy.

Mutual company: The policyholders own the company. Usually, a policyholder pays a small fee up front and during profitable years they may receive a dividend.

Named Insured: Person or entity named on an insurance contract that is therefore entitled to insurance benefits according to the provisions of the insurance policy. Also referred to as a policyholder.

Pedestrian: Includes a person on foot, a bicyclist, or motorcyclist.

Where coverage applies: Most policies written in the United States apply to accidents only in the United States. When traveling in other countries, be sure to check your policy because you will likely need additional coverage.

Read the definitions section of your policy to be clear about the meaning intended by your insurance policy.

COMMON LEGAL TERMS

Please note: For specific legal definitions, please review a good legal dictionary or discuss with an attorney. These comments are only designed to assist a layperson with understanding some basic legal concepts.

Burden of proof: in a criminal trial, the prosecution must show that the defendant was guilty beyond reasonable doubt. In a civil trial, such as those referred to in this text, only the preponderance of the evidence is needed to establish proof. In other words, there could be reasonable doubt about liability, but if most of the evidence supported the plaintiff's case, he or she could prevail.

Complaint: the basis for a lawsuit that spells out why the legal action is brought against the defendant.

Contingency fee: a fee owed based on whether or not a person is able to recover all or part of damages. In other words, it is an agreement between the plaintiff's attorney and his client that spells out what part of any award goes to the attorney. (Please note that even if no damages are awarded and no fee paid, there will still be costs associated with the lawsuit that the plaintiff must pay.)

Damages: the amount of loss specified in the lawsuit. Recoverable items are specified by law in each state and may not include items an ordinary person might view as damages.

Default judgment: when a defendant does not respond to legal proceedings, the plaintiff can get a judgment by virtue of the defendant's default.

Deposition: a recorded interview of any person who may testify in court, usually by a court reporter. All parties to the lawsuit may question this person under oath. This interview can be used as evidence in a trial.

First-party claim: one party seeks compensation from another based on their contractual agreement. (An insured and an insurer are the two parties to the insurance contract.)

Interrogatories: a set of written questions used by both plaintiff and defense to gain information.

Liability: responsibility for an accident as determined by applicable law. Generally, liability is established by analyzing the facts for evidence of negligence or violation of legal statutes.

Negligence: failure to use the reasonable care expected of an average person in a particular situation. This is a broad term and should be looked up in a legal dictionary. Several Web sites offer free definitions of legal terms.

Plaintiff: the person or party bringing the action in a lawsuit. The first listed name in the title of a legal proceeding.

Statute: a specific law written and passed by a legislature.

Summons: the first step in a lawsuit. It is a notice sent by the plaintiff to the defendant requiring a response or appearance in court by a certain date.

Third-party claim: a claim where a person or entity outside the contractual agreement may have a claim against either or both of the parties to the agreement. In the case of auto accidents, that third party may have a claim because the insured may be legally liable for the third party's damages. The important point is that the third party usually cannot sue the non-liable party, the insurer. Therefore, insurers don't worry when a third-party claimant threatens to sue them. They know they don't have to worry about their company, only their insured.

IF AN ACCIDENT OCCURS

Think personal safety first. Remove yourself or others from harm's way. Check for danger from traffic first. If there is danger, enlist any available help in keeping yourself and others safe. Try to remain calm.

Turn on your hazard lights.
Assess whether there are injuries. If so, call 9-1-1 immediately. Be prepared to give accurate information about the location. Then enlist any available help in rendering first aid. Do not attempt to move an injured party, unless absolutely necessary to prevent a life-threatening injury.

If you are injured, tell the 9-1-1 operator and then obtain the medical treatment you need.

If a nonemergency, dial 3-1-1 if you do not have a local number for the police. If the police are unable to come to the scene, obtain information on how to file a report at the local station or Department of Motor Vehicles. In this event, document the facts and damages as much as possible. Photographs and handwritten statements from witnesses will be helpful.

Get contact information from any witnesses immediately. This may be your only opportunity to obtain this information.

Take photographs of the vehicles for documentation of the accident. If you do not have a camera, check to see if your cell phone is capable of taking photographs.

If your vehicle is drivable and obstructing traffic, move it out of the roadway. If liability is questionable and you have no camera, you should ask police before moving your vehicle.

Keep calm and avoid any argument.

Do not sign any document other than for the police officer at the accident scene. Later, discuss with your insurance agent or attorney before signing any document.

Obtain insurance, vehicle registration, and identification information from the other driver. If the owner is someone other than the driver, obtain information about that person or organization as well.

Take notes on what happened, who was there, and where they were in relationship to the accident. Also note where the vehicles were located and the damage that occurred to them. Again, most cell phones have camera capabilities and can be used to document damage as well as spatial relationships.

Do not leave the scene of the accident until police get there, unless they advise you to do so. Most states have laws that require a person not to leave the accident location. Leaving the scene can carry a stiff penalty, especially if there are injuries.

If your car should be towed, remove all personal possessions from it.

Report the accident as soon as possible to your insurance agent. If you are driving a car owned by another person or entity, contact them immediately and ask them to contact their insurer. Generally, insurance coverage follows the car, and the insurer of the car will provide primary coverage. However, all potential insurers should be contacted right away or benefits could be jeopardized.

If the accident occurs in a parking lot, locate any possible witnesses immediately. Since liability is determined strictly by negligence in these situations, these accidents are always questionable. If there are no witnesses, you should check to see if the business has any security cameras that might have recorded the event. Document any skid marks, vehicle positions, location of damage, and severity.

DISPUTE RESOLUTION

How do you resolve a dispute as the result of an accident?

The first step to avoiding any dispute is to immediately report an accident to your own auto insurance company. Many people are reluctant to do this because they are afraid it will raise their insurance rates. This should not be a consideration. You have paid for this protection and should take action to get your claims handled. If the accident involves a third party, there may not be enough coverage to handle all claims. Therefore, you should take advantage of any first-party coverage that you have purchased.

Your claim will be assigned to an auto insurance claim representative. This is the first person with whom you should try to resolve a problem. If your insurer provides local claim service, try to meet with the claim representative in person. Face-to-face interactions can be a more effective way to communicate. Telephone, e-mail, or snail mail communication can lack courtesy, tone, or empathy and is devoid of any body language.

Many disputes result from poor communication. A helpful tool is to repeat what is said back to the person with whom you are communicating to ensure both parties clearly understand what is being communicated.

If you feel the claim representative is not handling your claims to your satisfaction, you can ask to speak with his or her supervisor. If you do not get satisfaction with the supervisor, then you can contact your state's insurance department and present a complaint. Most states have persons who investigate these complaints and will attempt to help resolve disputes.

Of course, you can hire an attorney if you become involved in a dispute. Many attorneys recommend that you hire an attorney right away. This can be helpful in order to preserve evidence. Skid marks fade, damaged cars get repaired or sold for salvage, witnesses can be difficult to locate, and memories of what happened can change or fade over time. Some attorneys employ accident investigators and in complex cases, an accident reconstructionist can be utilized.

Insurance companies will usually conduct their own investigations, but they can be inadequate because of the differing skills and biases of individual claim representatives. They are human and sometimes work with a heavy claim load.

The nature of the dispute may be helpful in deciding which route to take. If liability is disputed (an argument over who is at fault or liable for the accident), a good investigation will be important to the outcome. You can let your insurance company handle your claim and then pursue the claim against the other insurer. Most insurers are members of an Inter-company Arbitration Agreement and can submit their claims to an arbitration panel. If an insurer loses in arbitration, it will usually pay your disputed liability claims. However, your success may depend on how well your insurance company presents your claim and the level of expertise of the arbitration panel. The arbiters are usually composed of insurance company claim supervisors.

If you are not knowledgeable about the law or do not believe that your insurance claim representative did a good job with the investigation of the claim, then hiring a good attorney may be the best choice. Even though an attorney can be expensive, it may be worth it to establish liability, especially if you have a substantial amount of damages.

If liability is clear, then you probably will be able to handle your claim up until the point where the values may be disputed. At that point, if you decide to hire an attorney, you can negotiate his or her fee based on what he or she can get above and beyond what has been offered to you. Some may not accept such an arrangement, but there may be some who will.

Mediation is a method of dispute resolution where the parties to a dispute meet with a mediator who has experience in resolving auto accident disputes. Some are former judges and others have credentials based on experience. They will have their credentials available on request. Some require that the parties have legal representation.

Usually each side prepares their case in advance and sends support for their position to the mediator in advance. Then, the mediator meets with each side independently at the mediation proceeding. The mediator points out the merits and shortcomings of each position. Sometimes information will surface that helps resolve the dispute. In many cases, compromise can help settle disputes. Nothing that occurs in mediation can be binding on either party unless an agreement is reached.

Arbitration is a binding arrangement in which each side presents their case to the arbitrator, and he or she makes a decision, much like a trial. The advantage is that this can save you time and eliminate the need for an expensive, stressful, and time-consuming lawsuit. This form of dispute resolution helps reduce the number of lawsuits a judge has to hear. This method also eliminates any jury bias. Most arbitrators are former judges, and they usually require that each party has legal representation.

It has been said that in a trial, the only winners are the lawyers. That is not always the case. However, you should try to be objective about the evidence in your case and make your decisions accordingly. There are situations where going through a lawsuit is only way you can resolve a dispute.

TIPS FOR BUYING A CAR

New cars: Most automobile manufacturers have very helpful Web sites that allow you to investigate equipment options and prices for their cars. They also have dealer locations to assist you in finding a dealer close to you.

Decide what options you want and can afford *before* you shop. Don't be tempted to buy more once you get to the dealership.

Please note that once you arrive at a dealership you will find that prices are usually higher than those given on the Web site. This is due to add-ons such as pin stripes or undercoating which increase the dealer or distributor's profitability. If you elect to purchase a vehicle with these nonfactory additions, negotiation on price for these items is recommended.

Also, search the Internet for pricing information so that you will be knowledgeable about negotiating a price for the car.

If you are trading in a car, consider selling it yourself. Many dealers are aware that most people think the value their car is higher than its actual worth. They also know that many people may owe more on their car than it is worth. Therefore, they may offer a high trade-in price for your car but negotiate little on the price of their car. Chances are that you can sell your car for a better price than you'd actually get from a dealer. Many new car dealers simply wholesale cars traded in to used car dealers. Ask about deals both with your trade-in and without.

Consult the crash test Web sites to see how the car you select ranks in terms of safety. Some manufacturers make some safety features optional that may be included in the test car, so be sure the car you select has same features as the test car. If you can afford it, opt for as many safety features as possible, especially side curtain airbags.

Check the NADA (National Auto Dealers Association) guidebook to see how well the car you are thinking of buying will hold its value. There are also Web sites that claim to provide this information.

Check bank financing or credit union terms to compare with dealer financing. Sometimes the simple versus compounding of interest can make a difference.

A good rule of thumb for financing is to pay down at least 25 percent of the car's value. This will help keep you from owing more than the car is worth. If you are unable to do this, consider purchasing an auto insurance policy that will pick up the difference should you be involved in an accident. This is called "gap" insurance. More than one person has ended up in serious debt after wrecking a car that was purchased with no money down.

Used cars: You should use caution when purchasing a used car.

An important consideration is whether the car has been wrecked. A minor accident repair job by a reputable body shop should not cause serious alarm. But, if the car has been totaled by an insurance company and rebuilt then you should beware. Unethical rebuilders might cut corners on the repairs. For example, air bag components are very expensive to replace. To discover whether a car has been totaled, ask to see the title. Most of them will have the word "salvage" or "rebuilt" indicator on them. Carfax is another way to check the vehicle history. This service is available online and some dealers will check this themselves and make the information available to you.

Another way to check for structural damage is to carefully examine the lines where body parts meet. Any irregularity in the distance could be an indicator of improper repairs.

Mechanical problems are another area of concern. Ask if service records are available. If so, check the dates of service and the mileage at the time the maintenance was performed. You can hire a mechanic to inspect the car. You should discuss this with the mechanic beforehand to determine exactly what he will do and what problems he might be able to detect.

You definitely want to have good brakes. In addition to using the brakes while on a test drive, you can also roll down the window and listen as you apply the brakes. You should hear no noise. Brake wear can be determined by a mechanic.

A check of fluids in the car can be helpful. Most dealers are going to be sure that fluids are changed before placing the car on their lot. But, if you are buying from an individual, they may not. If the oil is a dark color, it is likely the owner does not change oil regularly. If you have little knowledge of auto mechanics, perhaps you can take a friend who does to help with identifying problems indicated by the condition of fluids. Also check for fluid leaks. When you move the car for a test drive, look at the area where it was parked. If you see an oil leak, you should discover the cost to fix the leak before buying the car.

Inspect the interior for abnormal wear in areas such as brake or clutch pedals. If the mileage is low, there should not be excessive wear.

If possible, try to find out how the vehicle was used. If you are buying a low-mileage vehicle from someone who used it on a farm or only drove a mile or two each day in city driving, the vehicle may have a shorter-than-expected life. For example, sludge may form in the oil pans of these vehicles and the engine may not get the lubrication it was designed for.

Discuss a warranty with the owner or dealer.

If the vehicle is more than four-years-old, it may behoove you to flush the radiator, replace the hoses and thermostat, and refill with new antifreeze. Some manufacturers recommend replacing antifreeze every year or two, but few people do that. Replacing belts and battery are also a good idea to prevent you from having a problem on the highway.

Search the Internet for additional car buying tips.

SAFETY SUPPLY CHECKLIST

Antibacterial hand cleaner: use this to keep your hands clean. Use to remove any gasoline or oil residue after filling your gas tank and checking oil level.

Blanket: keep at least one blanket in your car in the event you get stranded during winter weather or to use in the event of traumatic injury to prevent shock.

Coat: always keep a coat in your car when driving in winter weather. If traveling a distance during a snowstorm, carry food, water, boots, hat, gloves, and plenty of warm clothing.

Emergency phone numbers: 9-1-1 is the number to dial for emergency. 3-1-1 is a number for nonemergency services. To obtain information, use 4-1-1. In addition, keep numbers for family in your car in the event of an accident. Numbers for local wrecker services and your mechanic could be helpful in the event of a problem.

Emergency warning device: keep cones, fluorescent triangles, flares, or any other visible device to alert other motorists in the event your car becomes disabled.

Fire extinguisher: be sure you and your family understand how to use the fire extinguisher. Keep one appropriate for putting out engine fires in your vehicle. Read the instructions and find a safe location in your vehicle to store the device.

First-aid kit: familiarize yourself and other family members with the contents of the kit. Teach your children proper use of first-aid techniques. A first-aid class is recommended along with CPR training.

Flashlight: keep at least one flashlight in your car for problems that may occur while driving at night. Change the batteries at least once per year at the same time you replace batteries in your home smoke detectors.

List of medications: keep in your purse, wallet, or on your person in the event you are incapacitated.

Notification of any medical problem: if you have a medical problem that could complicate treatment of injuries, such as diabetes or mental condition, wear designated tags or bracelets to alert emergency personnel to your condition. If no such identification is available, write the information on a card and have it laminated.

Owner's manual: much valuable information is contained in these booklets.

Pen and paper: for recording information should an accident or problem occur.

Set of jumper cables: read instructions in your owner's manual so that you'll be prepared in the event your battery fails. The cables must be connected in the proper sequence and at the correct places. Failure to do so could result in injury.

Tire patch kit and small air compressor: for situations where the tire is deflated because of a slow leak, such as a nail or screw in the tire, an air compressor will reinflate the tire enough to get you to a tire dealer to have the tire repaired. If the flat is caused by a fast leak, you can plug the tire with a repair kit and reinflate. Then go to a tire dealer and have the tire inspected and/or repaired properly.

This book: bookmarked at the section What to Do in the Event of an Accident.

Umbrella: keep an umbrella or rain suit in your car for problems encountered when the weather may be rainy.

ARKANSAS WRONGFUL DEATH STATUTE

16-62-102. Wrongful death actions - Survival.

(a)(1) Whenever the death of a person or a viable fetus shall be caused by a wrongful act, neglect, or default and the act, neglect, or default is such as would have entitled the party injured to maintain an action and recover damages in respect thereof if death had not ensued, then and in every such case, the person or company or corporation that would have been liable if death had not ensued shall be liable to an action for damages, notwithstanding the death of the person or the viable fetus injured, and although the death may have been caused under such circumstances as amount in law to a felony.

(2) The cause of action created in this subsection shall survive the death of the person wrongfully causing the death of another and may be brought, maintained, or revived against the personal representatives of the person wrongfully causing the death of another.

(3) No person shall be liable under this subsection when the death of the fetus results from a legal abortion or from the fault of the pregnant woman carrying the fetus.

(b) Every action shall be brought by and in the name of the personal representative of the deceased person. If there is no personal representative, then the action shall be brought by the heirs at law of the deceased person.

(c)(1) Every action authorized by this section shall be commenced within three (3) years after the death of the person alleged to have been wrongfully killed.

(2) If a nonsuit is suffered, the action shall be brought within one (1) year from the date of the nonsuit without regard to the date of the death of the person alleged to have been wrongfully killed.

(d) The beneficiaries of the action created in this section are:

(1) The surviving spouse, children, father, mother, brothers, and sisters of the deceased person;

(2) Persons, regardless of age, standing in loco parentis to the deceased; and

(3) Persons, regardless of age, to whom the deceased stood in loco parentis at any time during the life of the deceased.

(e) No part of any recovery referred to in this section shall be subject to the debts of the deceased or become, in any way, a part of the assets of the estate of the deceased person.

(f)(1) The jury or the court, in cases tried without a jury, may fix such damages as will be fair and just compensation for pecuniary injuries, including a spouse's loss of the services and companionship of a deceased spouse and any mental anguish resulting from the death to the surviving spouse and beneficiaries of the deceased.

(2) When mental anguish is claimed as a measure of damages under this section, mental anguish will include grief normally associated with the loss of a loved one.

(g) The judge of the court in which the claim or cause of action for wrongful death is tried or is submitted for approval of a compromise settlement, by judgment or order and upon the evidence presented during trial or in connection with any submission for approval of a compromise settlement, shall fix the share of each beneficiary, and distribution shall be made accordingly. However, in any action for wrongful death submitted to a jury, the jury shall make the apportionment at the request of any beneficiary or party.

(h) Nothing in this section shall limit or affect the right of circuit courts having jurisdiction to approve or authorize settlement of claims or causes of action for wrongful death, but the circuit courts shall consider the best interests of all the beneficiaries under this section and not merely the best interest of the widow and next of kin as now provided by § 28-49-104.

(i) It is not the responsibility of the personal representative of a deceased person to locate anyone in loco parentis who is not known to the personal representative to be in loco parentis to the deceased person.

History. Acts 1957, No. 255, §§ 1-5; 1981, No. 625, § 1; A.S.A. 1947, §§ 27-906 - 27-910; Acts 1993, No. 589, § 1; 2001, No. 1265, § 1; 2001, No. 1581, §§ 1, 2.

AMI 2216 (The following are instructions to be used by a judge for instructing a jury hearing a death claim.)

Measure of Damages-Wrongful Death-Cause of Action

as administrator of the estate of, deceased, represents the estate of the deceased and also (names of wife or husband, children, father, mother, brother, sisters or persons in loco parentis for whom claims are made) The administrator is suing for the following elements of damage on behalf of (wife or husband) [and] (names of statutory beneficiaries)

(a) [Pecuniary injuries sustained by ( ) (and)
(wife or husband) (names of statutory beneficiaries entitled to recover for pecuniary injuries)

(b) [Mental anguish suffered (and reasonably probable to be suffered in the future) by ( ) (and) ( )] (wife or husband) (names of statutory beneficiaries making claim)

(c) [Loss of consortium of].(wife or husband)

First, let me explain to you what is meant by the term "pecuniary injuries." This term refers to the present value of benefits, including money, goods, and services, that the deceased would have contributed to [and] had he lived. In making (names of appropriate statutory beneficiaries) your determination of pecuniary injuries, you may consider the following factors concerning the deceased:

(a) [What he customarily contributed in the past and might have been reasonably expected to contribute had he lived.]

(b) [The period during which any beneficiary might reasonably expect to have received contributions from the deceased.]

(c) [What the deceased earned and might have been reasonably expected to earn in the future.]

(d) [What he spent for customary personal expenses and other deductions.]

(e) [What instruction, moral training, and supervision of education he might have reasonably given his (child) (children) had he lived.]

(f) [His health.]

(g) [His habits of industry, sobriety, and thrift.]

(h) [His occupation.]

(i) [The life expectancy of the deceased and of (wife or husband) (and) (names of appropriate statutory beneficiaries)

(j) [The time that will elapse before his (child) (children) reach(es) majority.]

Second, let me explain to you what is meant by mental anguish. This term means the mental suffering resulting from emotions, such as grief and despair associated with the loss of a loved one.

Third, let me explain what is meant by the term "consortium." Consortium refers to the society, services, companionship, and marriage relationship of the [husband] [wife].

The administrator is also suing for the following elements of damage on behalf of the estate:

(a) [The reasonable value of funeral expenses];

(b) [Property damage];

(c) [Conscious pain and suffering of the deceased prior to this death];

(d) [Medical expenses attributable to the fatal injury];

(e) [The value of any (earnings) (profits) (salary) (working time) lost by the deceased person prior to this death];

(f) [Any (scars) (disfigurement) (and) (visible results of the injury) sustained by the deceased prior to his death];

(g) [The reasonable expenses of any necessary help in his home that prior to his death was required as a result of the deceased's injuries].

[If you decide for the administrator on the question of liability (against any party he is suing)] [If an interrogatory requires you to assess the damages of the administrator], you must fix the amount of money that will reasonably and fairly compensate [ ] [and] (wife or husband) [ ] [and] [the estate] (names of other statutory beneficiaries making claims) for those elements of damage you find were proximately caused by the [negligence] [or] [fault] of Whether any of the damages sued for on behalf of [ ] [and] (wife or husband) [ ] have been proved by (names of statutory beneficiaries making claims) the evidence is for you to determine.

POTENTIAL PITFALLS IN AN AUTO INSURANCE POLICY CONTRACT

WARNING: The comments in this section are designed only as an aid to understanding the auto insurance policy. Since there are variations between the policies issued by different companies, you should review your policy carefully and obtain the opinion of an attorney if you want to be sure the risks in your activities are covered by your policy contract. The examples listed in this section are only some of the situations that may not be covered.

A Typical Auto Insurance Policy Contract

The next few pages explain a typical auto insurance policy in easy-to-read terminology. An insurance policy is the contractual agreement between you and your insurance company. You pay your money (they call this premium) in exchange for their promise to pay according to an agreement similar to this.

Declarations Page

The declarations page tells who is insured and what vehicle is covered by the agreement. Please note that under certain circumstances, a vehicle not listed here and drivers not listed on the declaration page can qualify for coverage.

The declarations also tell you what coverage the policy provides (what type losses they will pay for) and the amounts payable. Carefully read the amounts listed under the liability, medical payments, uninsured, and underinsured coverage. These are of critical importance. You should also examine amounts payable under total disability, death and dismemberment coverage. To find that on some policies you'll have to look at the policy contract.

Most policies specify that you own the vehicle described. Be sure and tell your insurer if this is not the case, so that you can purchase the appropriate policy for protection.

You should also tell your insurer if you or a family member has had insurance canceled or not renewed by another company in the past three years.

Also, if you or any member of your family (that lives with you) has had a license or vehicle registration suspended, revoked, or refused during that three-year time period, you should tell your insurer. If you don't, your insurer could have your policy annulled, which means the agreement cannot be enforced. In other words, you may not get paid for a loss. This clause is contained in the policy section of your agreement.

Definitions

This section of your policy defines the terms used. Some common definitions follow:

Bodily injury - must be to a person, not an animal such as your prized show dog, and only includes sickness, disease, and death. The purpose of this is to avoid other types of injury like libel.

Car - yes, you know what a car is, but the insurer defines it so as to avoid covering a house, trailer, semi tractor, etc.

Newly Acquired Car - this tells you the number of days you have protection for a newly purchased car. To be safe, you should contact your insurer right away and be sure you have protection. My policy provides twenty days protection after I take delivery of a new car.

Nonowned Car - This tells you the circumstances where cars that you don't own are covered by your policy. There is not a lot of protection, so if you are using a car that you do not own, be sure that it is listed on an insurance policy by the owner and that you are covered as a driver. Most policies cover any driver who has permission from the owner. Some notable exceptions are people in a car business such as a taxi service or a mechanic. See the exclusions section of your policy.

Occupying - this tells you where in relation to your car that an incident will be covered. More on this later.

Most definitions are easy to read and relatively straightforward. You will need to refer to the definitions section frequently to understand the specific provisions under each coverage.

Liability Section

This section begins by telling you who is insured and what cars are covered under your liability coverage.

Insuring agreement - this promises to pay for the damages you cause to others in a car wreck. This has to be accidental. If you intentionally injure someone with your car, you may have no protection under your policy. For example, if a person becomes enraged and causes injury to another driver because of an intentional act, there may be no coverage. Therefore, you should keep this in mind if you loan your car to someone who may be prone to anger. What can happen is that this type of incident will be considered a coverage question. In severe cases, your insurance company may file a lawsuit asking a judge to rule on the coverage question. In that event, you could be faced with paying for an attorney to defend you in that action. If the insurance company wins, you will be personally liable for any damages if the injured party gets a judgment against you.

Generally speaking, you must be driving a car that is covered in your policy agreement. However, you should report any accident to your insurer involving you or any family member who resides with you in case there may be additional protection available.

The insuring agreement also says it will pay attorney fees for your defense if you are sued by a third party. Most only agree to pay fees to attorneys the insurance company selects, so talk to them before hiring an attorney to defend you. Also, if you do happen to get sued, please read this section very carefully so you'll know exactly what they will pay.

The liability coverage section lists exclusions. These are claims your insurance company will not pay. The reasoning behind these exclusions is that most of them are covered by some other form of primary insurance. An insurance company can offer a lower premium by not covering these situations. Look at your risks carefully as you read the exclusions below and see if you may need some other form of liability insurance.

Household exclusion - most policies contain this clause and will not pay for injury under the liability coverage to family members who live with you. However, they are generally covered under the medical payments, total disability and death coverage, although this coverage is minimal.

If you want to protect yourself and your family, be sure to carry adequate health, life, disability, and long term care insurance so everyone in your family will be protected in the event of any injury. Your auto insurance will not pay for their injuries except as explained in the medical payments, total disability, uninsured motorist, underinsured motorist, and death and dismemberment coverage.

Worker's Compensation Exclusion - your auto insurer will not pay damages under the liability coverage that were intended to be covered by Worker's Compensation. If you have employees or travel with fellow employees in your car, you should read this section of your agreement very carefully, and then purchase any protection you need.

Personal Property Exclusion - your insurer will not pay for damage to property owned by you or any family member residing with you under this coverage. You probably already have your house, boat, cars, etc. protected. The home owner's policy has provision for damage to personal property, but most have a deductible and limitations. If you want your property protected be sure to get the physical damage coverage you need. For example, if you have an expensive computer for specialized work, you may want to buy extra coverage in the event it is damaged in an automobile accident. Your insurance agent will help you obtain extra coverage.

Assumed Liability Exclusion - you can't simply think you are at fault (liable), enter into an agreement to pay someone, and make your insurer pay the claim. Always report a potential liability claim to your insurer as soon as possible and ask them to talk about it with you. Most policies require you to report immediately.

Exclusion for Restitution in a Criminal Proceeding - your insurer will not pay for damages resulting from crime.

Exclusion for Off-Road Use - vehicles used off road for racing, climbing, etc. or on a race track will not be covered under the liability coverage. You can imagine the difficulty in proving fault for the damage that occurs in a dirt track car race. If you race cars, motorcycles, or any other motorized vehicle and want liability coverage, you'll probably need specialized coverage, or a legal document that protects you against liability from others.

Federal Tort Claim Act Exclusion - if you are an employee of the government and provisions of the Federal Tort Claim Act apply, you are not covered under the liability coverage. If you are an employee of the government or one of its agencies, please read this provision carefully and discuss it with your employer, your insurer; if there is any question about your protection, consult an attorney.

There are other exclusions in some polices, and you should read over them to see how they may affect you.

Another part of the liability section of your policy specifies what your insurer will pay if other liability coverage applies. This section should not affect you because you will have protection.

However, be sure to check that your policy has a provision that will provide you with the minimum liability protection required by law in other states. This is really important to those who live in states with low minimum liability insurance requirements such as Oklahoma.

Medical Payments Coverage

This section begins with definitions of who is insured and what is covered. Generally speaking, you and your family members who live with you are covered, and your insurance company will pay for reasonable medical services required as a result of an automobile accident. There can be limited protection for other people riding with you as well as pedestrians. This can vary by state. Check your agreement with your insurer for specifics.

In some states, this medical payments coverage is referred to as no-fault coverage because your treatment costs for an injury sustained in an accident are paid, regardless of liability.

Most policies have both a dollar amount and a time limit for treatment.

If you have this coverage and begin treatment for injuries you received in an automobile accident, check your policy for the time and dollar limitation. Plan accordingly.

Medical payments coverage can also have a provision for funeral expense.

If you or a family member dies as a result of an automobile accident, and there is provision under your medical payment coverage, you may want your medical expenses paid by your health carrier so that more of the funeral expense can be paid.

However, most health carriers take the position that all medical expenses caused by an automobile accident should be paid by the auto insurance carrier, the primary carrier. After the limits of the auto insurer have been exhausted, then they will apply the health policy, including deductibles, co-pays, etc. Their position is that this keeps health insurance premiums lower. An opposing argument is that you've paid two premiums and should get maximum benefit from both.

Medical payments coverage also contains exclusions. It contains a Worker's Compensation exclusion similar to the one in the liability section. A second exclusion prevents you from getting paid medical benefits from more than one policy (on other cars) for an injury in one accident.

Most policies also have a car-business exclusion. If you have a car business, you should read those exclusions carefully and consult a reputable agent and/or attorney.

Some policies exclude coverage for an insured while valet parking.

Another exclusion can prevent you from collecting for an injury that happened while you were occupying or struck by a camper trailer or recreational type vehicle used as a residence.

There is an exclusion for injuries you receive if you are walking or riding a bicycle or motorcycle off public roads and are struck by a vehicle designed for off-road use such as a dirt bike, train, or bulldozer.

Most policies exclude medical payments coverage for injuries from war, nuclear problems, gunshot, fungi, racing events, intentional injury, committing a felony, or while running from police.

Total Disability Coverage

This provides payment if you are unable to do your job while you are continuously and totally disabled. They will not pay for partial disability. Usually there is an upfront waiting period. For my insurer, that period is eight days. You should check your policy.

The amount you receive depends on the limit you choose and whether you are an income producer or not. You should read your policy carefully for the limit and whether it meets your needs when you purchase insurance. My insurer has three options: $140, $250, or $500 as maximum amounts. Most also limit the time period. My insurer offers coverage periods of fifty-two weeks or one hundred and four weeks.

This coverage also contains exclusions.

Emergency vehicle exclusion - there is no coverage under the total disability coverage if you are injured in an emergency vehicle such as an ambulance, and your job is as the ambulance driver.

Business exclusion - if you're doing your job in a vehicle that is not a private passenger car. An example of that might be a truck driver.

Both medical payments coverage and total disability coverage have other insurance provisions that keep you from collecting twice for the same event. It is also designed to prevent "stacking" of insurance policies in the event you have more than one auto insurance policy in force.

Death, Dismemberment, and Loss of Sight

If you or a member of your family who lives with you dies, loses a limb, (such as arm, leg, finger, etc.) or eyesight, your insurer will pay according to a schedule listed in your policy. My insurer offers two choices for each of these benefits. Read this schedule carefully before purchasing your coverage.

Some policies will double the benefit if the injury occurs while the insured is properly wearing a seat belt in a private passenger car.

There are provisions for others outside your immediate family under certain circumstances. For example, my policy will provide benefits to a pedestrian, bicyclist, or motorcyclist involved in an accident with me, if the person does not have this coverage.

In my state, insurers are required by law to do this. Check your policy carefully before making any assumptions about who might be covered.

This coverage also contains exclusions.

Suicide exclusion - you cannot commit suicide and expect your insurer to pay benefits. In fact, even if you were insane and attempted suicide, my insurer would not pay you. Insurance was never intended to pay for intentional, destructive acts of an insured.

Disease - Automobile insurance will not pay for death, dismemberment, loss of sight, or total disability that is caused by disease, except for infection that results from injury sustained in the accident. There are other policies to cover those risks.

The exclusions that apply to the medical payments coverage also apply to the death, dismemberment, and loss of sight coverage. Please refer back to that section. Again, read your policy because there is variation among insurers.

Uninsured Motorist Coverage

This is an important coverage that I strongly urge you to consider. In spite of the fact that most states require insurance, many people find ways to cheat the system.

If you do not carry uninsured motorist coverage, then you'll only be paid by your own carrier for the protection you've agreed to. That means you'll have to pay a deductible for repair or replacement of your car and only have medical payments coverage, total disability, death, dismemberment, and loss of sight protection for any injury. These first party coverages offer fair protection for minor injury but are inadequate for severe injury. Uninsured motorist coverage is a relatively inexpensive way to increase your protection.

Often, people who do not have insurance are dangerous drivers. In my opinion, they pose a greater risk of injury to you than the average driver.

Therefore, I recommend that you buy uninsured motorist coverage. This pays you for property damage, medical expenses, income loss, car rental, (depending on how your state interprets loss of use damages), plus pain and suffering. In short, your insurer pays just as if the uninsured driver had liability insurance coverage.

This coverage applies to you and your family members who live with you. In some circumstances, others are covered, but please read your policy, especially if you have an accident with an uninsured motorist.

If you are involved in an accident in another state or while in another car, you should report your claim to your insurance company. Benefits can be altered by differing state law or policies from different insurance companies.

The exclusions listed are usually similar to those in the liability section as well as other parts of your agreement.

If an uninsured motorist offers to pay you for your damages, you must discuss the offer with your insurer. If you settle your claim with the uninsured driver, you may not be able to collect from your insurer if problems develop later. Some insurers require that you get their written consent before accepting a settlement offer from an uninsured motorist. They do this so that the uninsured can be held responsible for damage they cause. If your insurer does not consent to the agreement, they will likely go ahead and pay you what the uninsured driver has offered and wait to make sure that you have no further problems before concluding your claim.

Similarly, your insurer will not be bound by a judgment against an uninsured driver unless you have followed the insurance policy agreement. The reason is that sometimes a default judgment can easily be obtained against an uninsured driver because he or she may not follow proper legal process. Read your agreement and discuss with your insurer and/or an attorney.

Underinsured Motorist Coverage

In most states, the minimum liability coverage required by law is $25,000 for property damage (for repair or replacing cars and other property) and $25,000 per person for paying injury claims, with a $50,000 cap for all claims, regardless of how many exist. Presently, that is not enough money for many claims. Therefore, if you want good protection for severe injury that could be caused by others, this coverage is very important, and you can select from several amounts.

Most of the provisions from the uninsured motorist coverage also apply to underinsured motorist coverage.

Physical Damage Coverage

This section pays for damage to your car and, under some circumstances, can pay for damage to a car you borrow.

Typically a policy has collision coverage that pays for repair or replacement of your car due to a collision. Your insurer will subtract your deductible from any payment.

If your policy specifies use of "like kind and quality parts" then your insurer can list salvage parts or aftermarket parts for the repairs to your car. You will be expected to pay the difference if you want new original equipment manufacturer parts used for repair of your vehicle.

Comprehensive coverage pays for damages that result from other causes such as fire, hail, theft, etc. This coverage is very broad and includes unusual circumstances, such as damage caused by animals. If you have a loss that is not due to ordinary wear, ask your insurer whether it might be covered under comprehensive coverage. Again, your deductible will be subtracted from any amount paid.

Towing coverage will pay for a tow if your car is disabled.

Car rental coverage pays for a rental car when your car is disabled by a loss covered in this section of your policy. Most do not pay all the rental costs. Read your policy to determine the limits of this coverage.

This coverage would not pay for a rental car while the insured vehicle is being repaired because of something not covered by the insurance policy such as an engine overhaul.

There is no coverage for damage you intentionally cause.

Similarly, there is no coverage for racing, or if a government authority seizes your car.

There is no coverage for ordinary wear and tear, freezing, or mechanical breakdown.

Read your policy carefully for exclusions. In a typical auto insurance policy there are nineteen descriptions for situations in which the policy would not apply. It is your responsibility to read and understand your insurance policy contract.

When your children move out of the household and are not in college you should be sure they are listed on their auto insurance policy as the named insured because the protection will be greater.

SAMPLE LETTER

Date

Name
Address

Dear (Elected Representative)

As my elected representative, I would like for you to use your power and influence to reduce the numbers of automobile accidents on our roadways and to take measures that would reduce the economic consequences of those accidents that do occur. During the past twenty years, over six million accidents have occurred each year, with more than two million persons injured, and approximately forty thousand killed each year. This is a leading cause of death for our young citizens.

Automobile accidents are caused by the following driver behaviors: aggressive driving, following too closely, speeding, inattentive driving (such as using cell phones), impaired sensory ability, and reckless driving.

Because many people persist in these behaviors, enforcement of law prohibiting these behaviors must be improved. The most effective way to achieve this would be to increase police staffing, provide automated traffic monitoring, and place video equipment in all police cars.

Although most schools include driver's education in their curriculum, please incorporate graduated licensing for youthful drivers on a national level.

Similarly, please implement a program for testing of senior drivers based on anonymous reports. As baby boomers age and begin to experience the natural effects of aging, such as loss of sight, hearing, or mental abilities, the risk of accident and death increase. Mandatory testing may place an undue burden on many drivers, but at least those who are at risk should be tested.

Other prevention measures include: prohibiting nonemergency personnel from driving during times of extremely hazardous roadway conditions; developing safety law for roadway and business zoning design; improving the regulation of the trucking industry with regard to traffic safety; decreasing speed limits on heavily traveled interstates, which has been shown to reduce fatalities and dependence on foreign oil; and improving emphasis on safety in regard to regulations on automobile manufacturers.

These prevention techniques could be handled by the National Highway Traffic Safety Administration and the Department of Homeland Security.

In order to reduce the consequences of automobile accidents, I ask you to enact an undiluted no-fault, pay-at-the-pump, government-controlled plan that provides only compensatory benefits. Those who participate could not sue or be sued as a result of an automobile accident. This law should be specific about benefits, and those handling claims should have similar legal powers to those implemented for Medicare.

Please work with the medical community to make health care more affordable and accessible for consumers.

Those who do not wish to participate can continue to use private plans.

In the event you do not wish to enact no-fault legislation, I ask that you enact tort reform so that if plaintiffs and their attorneys fail to obtain a verdict larger than the amount offered by the defendants, they must pay all costs of the defendant, including attorney fees, court costs, and the amount of the award. Conversely, if the plaintiffs obtain a verdict larger than what the defendants have offered, the defendants and their legal representatives must pay those costs.

For more information on these subjects, I suggest you read a book titled *American Highway Roulette*.

Please represent my interests as a taxpayer, consumer, and voter in the above matters.

Sincerely,

(Your name)

WEB SITES FOR INFORMATION

AAA Foundation for Traffic Safety Web site: www.aaafoundation.org

Advocates for Highway and Auto Safety Web site: www.saferoads.org

American Association of State Highway and Transportation Officials Web site: www.transportation.org

American Baptist Churches USA Web site: www.abc-usa.org and link to the Resolution on Highway Safety: www.abc-usa.org/resources/resol/highsafe.htm

American Driver and Traffic Safety Education Association Web site: www.adtsea.1up.edu

American Highway Users Alliance Web site: www.highways.org. This Web site claims to be the united voice of the transportation community.

American Public Transportation Association Web site: www.apta.com

American Tort Reform Association Web site: www.atra.org

American Traffic Safety Services Association Web site: www.atssa.com

American Trucking Associations Web site: www.truckline.com

Association of Industrial Road Safety Officers (United Kingdom organization) Web site: www.airso.org.uk

Association of Trial Lawyers of America Web site: www.atlanet.org

Car Talk NPR Radio Talk Show Web site: www.cartalk.com

Citizens Against Speeding and Aggressive Driving Web site: www.geocities.com

Coalition Against Insurance Fraud Web site: www.insurancefraud.org

Concerned Americans for Responsible Driving, Inc. Web site: www.drivingsafe.org (Web site for story about senior drivers)

Consumers for Auto Reliability and Safety Web site: www.carconsumer.com

Continental Teves Web site: www.contiteves-na.com

Friends of the Earth Web site: www.suv.org

Governors Highway Safety Association Web site: www.naghsr.org

Institute for Legal Reform Web site: www.instituteforlegalreform.com

Insurance Information Institute Web site: www.iii.org/

Insurance Institute for Highway Safety Web site: www.iihs.org/

Insurance Research Council Web site: www.ircweb.org

Keep Kids Alive Drive 25 Web site: www.keepkidsalivedrive25.org

Legal Ethics and Reform Web site: www.legalethicsandreform.com

Mothers Against Drunk Driving Web site: www.madd.org

Motorcycle Safety Foundation Web site: http://msf-usa.org

National Conference of State Legislators Web site: www.ncsi.org

National Health Care Anti-Fraud Association Web site: www.nhcaa.org

National Motorist Association Web site: www.motorists.org
Some opposing viewpoints can be found here.

National Sleep Foundation Web site: www.sleepfoundation.org and www.drowsydriving.org

Network of Employers for Traffic Safety Web site: www.trafficsafety.org

Operation Lifesaver Web site: www.oli.org

Power of Attorneys Web site: www.power-of-attorneys.com

Public Citizen Web site: www.citizen.org

Safe Ride News Web site: www.saferidenews.com

Southern Baptist Convention Web site: www.sbc.net and the link to their Resolutions is: www.sbc.net/resolutions/amResolution.asp?ID=599

The Center for Auto Safety Web site: www.autosafety.org

The NOLO Web site contains a wealth of legal information www.nolo.com

Truck Safety Coalition Web site: www.trucksafety.org

Tween Traffic Safety Web site: www.tweensafety.org

United States Department of Transportation, National Highway Traffic Safety Association (NHTSA) Web site: www.nhtsa.dot.gov provides information on vehicles, equipment, defects, traffic safety, research, and more.
NHTSA crash ratings can be found on www.safercar.gov

Web site for drivers, driving, and traffic safety: www.drivers.com

INDEX

AAA Foundation, 35, 145
Accident, what to do, 119
Addict, 82
Aggressive driving, 74, 80, 111
American Association for Justice, 87
American Baptist Resolution On Highway Safety, 80, 145, 147
American Insurance Institute, 21
American Tort Reform Association, 87, 145
Arbitration, 86, 123
Arkansas Wrongful Death Statute 129
Assigned risk pool, 92
Automated enforcement, 74
Average jury verdicts, 47
Bad faith, 26
Betterment, 115
Burden of proof, 117
Buying a car, 124
Capital stock company, 115
Car rental coverage, 142
Casinos, 4

Cell phones, 82, 110
Civil law, 42
Claim values, 15, 44
Claimant, 115
Coalition Against Insurance Fraud, 24, 25, 145
Collision coverage, 141
Comparative negligence, 47
Complaint, 45, 121
Comprehensive coverage, 142
Concerned Americans for Responsible Driving, 36, 146
Consumer, 5, 30, 50, 57, 60, 76, 79, 85, 88, 92, 94
Contingency fee, 117
Coverage, 133
Criminal law, 42
Damage disputes, 23, 43, 44, 122
Damage issues, 23, 43
Death, 97, 129
Death, dismemberment, and loss of sight, 139
Declarations page, 133
Deductible, 115

Default judgment, 141
Deposition, 117
Derivative claims, 17
Destiny, 98
Diminished value, 23
Dispute resolution, 121
Driving behavior, 79, 80, 110, 111
Driving behavior modification tips for parents, 107
Drowsy driving, 80
Economic costs, 79
Electronic stability control, 58
Evaluation, 15
Exclusions, 135, 136, 142
Fatalities, 2, 7, 33, 53, 71
Fraud, 24, 25, 67, 89
Federal Tort Claim Act Exclusion, 136
First party claims, 16, 118
Funeral expense, 27
God, 97
Government, 73, 85, 93
Government casino, 73
Graduated licensing, 36
Health care casino, 65
Health care challenge, 68
Health care private plans, 90
Health care public plans, 90
Health care fraud, 67
Highway Traffic Safety Administration, 1, 85
Indemnity, 115
Injury claim, 44
Institute for Legal Reform, 146
Insurance, 11, 115
Insurance fraud, 25
Insurance average annual cost 21
Insurance challenge, 29
Insurance employees, 21
Insurance minimum requirements, 19
Insurance problems, 22
Insurance buying tips, 114
Insurance casino, 11
Insurance Information Institute, 21, 146
Insurance Institute for Highway Safety, 22, 146
Insurance policy, assumed liability exclusion, 136
Insurance policy, Declarations Page, 133
Insurance policy, definitions, 134
Insurance policy, household exclusions, 135
Insurance policy, liability coverage, 134
Insurance policy, other exclusions, 136, 138, 139, 142
Insurance policy, personal property exclusion, 136
Insurance policy, worker's compensation exclusion, 136, 140
Insurance policy, potential pitfalls, 133
Insurance Research Council, 24, 146
Insured, 115
Interrogatories, 118
Legal Casino, 4, 35
Legal discovery, 45
Legal Issues, 47
Legal liability, 42, 47, 85
Legal process, 45, 47

Legal challenge, 50
Legal system strengths, 49
Legal system weaknesses, 49
Liability coverage, 118, 134
Licensed drivers, 79
Lobbying, 67, 90
MADD, 47
Mediation, 122
Medical payments coverage, 137
Medicare, 89
Motorcycles, 60, 116, 137
Mutual company, 115
National Health Care Anti-Fraud Association, 67, 146
National Highway Traffic Safety Administration, 1, 2, 79, 147
National Motorist Association, 91, 146
National Safety Council, 79
Negligence, 118
Newly-acquired car, 134
No-fault law, 25, 85
Non-owned car, 134
Pay at the pump, 84
Pedestrian, 116
Physical damage coverage, 141
Plaintiff, 118
Prevention of accidents, 79, 101
Private sector, 89, 90
Proactive law enforcement, 73, 80
Problems in politics, 75, 93

Public health care, 88
Punitive damages, 26
Reactive law enforcement, 73
Reducing frivolous lawsuits, 87
Runaway juries, 25, 88
Safe driving tips, 102
Safety supply checklist, 127
Sample letter, 143
Senior drivers, 35, 83
Sensory impairment, 82
Sobriety checkpoints, 82
Speed limits, 43, 48, 77
Statute, 47
Subrogation, 85
Summary, 101
Summons, 118
SUVs, 58
Third party claim, 15, 118
Total disability coverage, 138
Towing coverage, 142
Traffic safety, 102, 145-147
Trucking industry, 84
Underinsured motorist coverage, 76, 141
Uninsurable risks, 21
Uninsured motorist coverage, 48, 76, 140
Victim's fund, (9/11) 3
Web sites for information, 145
Where coverage applies, 116
Whiplash injury, 44, 66
Work hardening, 90
Youthful drivers, 36, 83, 107

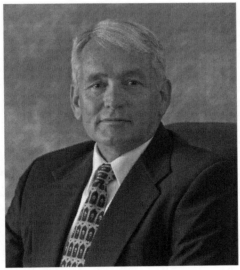

About the author: Denton Gay graduated from the University of Arkansas. In 1985 he began a career in automobile insurance claims. He completed the General Insurance curriculum, received an Associate In Claims, and earned a Chartered Property and Casualty Underwriter designation. He has sixteen years' experience in auto insurance claims with a large insurance company. He lives, works, and plays in and around the streams, lakes, and racquetball courts of northwest Arkansas. You may contact him at his Web site, AmericanHighwayRoulette.com.

Made in the USA
Charleston, SC
17 November 2011